Edson Alves Filho

Geoindicators of Environmental Change - Volume III

Edson Alves Filho

Geoindicators of Environmental Change - Volume III

A Reading of the Historical Geography of the Landscape - PCH Rio do Peixe I and II (1925-2016)

ScienciaScripts

Imprint

Any brand names and product names mentioned in this book are subject to trademark, brand or patent protection and are trademarks or registered trademarks of their respective holders. The use of brand names, product names, common names, trade names, product descriptions etc. even without a particular marking in this work is in no way to be construed to mean that such names may be regarded as unrestricted in respect of trademark and brand protection legislation and could thus be used by anyone.

Cover image: www.ingimage.com

This book is a translation from the original published under ISBN 978-613-9-64854-2.

Publisher:
Sciencia Scripts
is a trademark of
Dodo Books Indian Ocean Ltd. and OmniScriptum S.R.L publishing group

120 High Road, East Finchley, London, N2 9ED, United Kingdom
Str. Armeneasca 28/1, office 1, Chisinau MD-2012, Republic of Moldova, Europe
Printed at: see last page
ISBN: 978-620-7-79842-1

I would like to thank my supervisor Prof.ª . Drª . Sueli Angelo Furlan for conceiving and supporting this publication, and to my family for supporting my studies.

Summary

CHAPTER 1

PRESENTATION

This work is the book version of the Master's Thesis entitled "Geoindicators of Morphological Changes in Physical Systems Impacted by Hydroelectric Projects: A Reading of the Historical Geography of the Landscape - PCH Rio do Peixe I and II (1925-2016), presented to the Postgraduate Programme in Physical Geography at the University of São Paulo in 2017.

As you can see from the title of the book, there has been a slight change from the title originally published in the master's degree, replacing the term 'geoindicators of morphological changes' with the term 'geoindicators of environmental changes'.

This change is justified by the fact that morphological alterations can be read as changes caused by human action on the relief, with direct repercussions on the alteration of terrain features, which therefore corroborate environmental changes. In this way, a term that is very specific to the field of Geomorphology is no longer used, but rather a synonym that is more widely used outside the specialised field.

Due to the large amount of work done throughout the dissertation, it was decided to divide the work into three volumes. In Volume I, the objectives, premises, theme and justification of the research were presented, showing the importance of research within the geographical ternary into the field of short-cycle environmental changes by means of geoindicators and how the scarcity of research that parameterises these changes within the intervention modality of hydroelectric projects is still present, which justifies the pioneering nature of the work (Sections 2, 3 and 4).

Section 5 provides a bibliographical review of the concepts that underpin the research, centred on terms such as landscape, anthropogenic derivations, the electricity sector and anthropogenic interventions. Wide-ranging databases were used, such as Web of Science, the Digital Library Platforms of Brazilian public universities, various scientific articles and reference books on the topics considered.

Finally, Section 6 provides a detailed characterisation of the study area, through consultation of theses and dissertations on the region, field surveys and mapping using geographic information systems, giving a good overview of the various environmental aspects involved. Finally, in order to synthesise all the environmental characteristics of the study area, landscape units were proposed, which resulted in the creation of synthetic compartments of natural and man-made arrangements.

Section 2 of **Volume II of** the book provides an exhaustive theoretical review that will guide all the methodological procedures used in the research, covering theories and reference authors on the concept of landscape in Geography, passing through the notion of changes in environmental systems, until the founding theoretical framework of the research is defined, centred on authors and theories on anthropogeomorphology and impacts on the physical environment, the relief as a support for the landscape and finally, Detailed Geomorphological Cartography as a tool for assessing negative impacts on hydroelectric projects.

Section 3 provides an exhaustive inventory of all the methodological procedures used to achieve the research objectives, both in relation to the Integrated Environmental Assessment of the Physical Environment proposed for the study area, as well as the research and application of geo-indicators of environmental change in the phases of semi-preserved original morphology and anthropogenic morphology.

In this **Volume III**, more precisely in **Section 2,** all the results obtained will be presented, both those linked to the Integrated Environmental Assessment of the Physical Environment for the area directly affected by the Rio do Peixe I and II SHPs, and for the geo-indicators listed for the situations of semi-preserved original morphology in the pre-intervention phase, and for anthropogenic morphology in the active intervention and consolidated intervention phases.

To conclude this section, a comparison will be made between the official methodology of the Brazilian electricity sector for the environmental viability of hydroelectric projects and the methodology of geo-indicators of environmental change opened up by the anthropogeomorphological approach.

Finally, Section 3 presents some final considerations based on the results of the research and suggests new research topics and methodological paths for further study.

CHAPTER 2

RESULTS OBTAINED

1.1 Integrated Environmental Assessment of the Physical Environment in the Study Area according to the

Methodology of the Hydroelectric Inventory Manual (Eletrobrás, 2007)

As defined in the Methodological Procedures (Section 3.3, Volume II), the preparation of the Integrated Environmental Assessment of the Physical Environment in the Study Area involved the creation of a Physical Environment Sensitivity Synthesis Map (Map **2.1.l of** this volume), a Map of Negative Environmental Impacts on the Physical Environment (Map **2.1.m of** this volume) and, finally, a Physical Environment Fragility Map **(Map 2.1.n of this** volume).

It should be emphasised, however, that the preparation of the Synthesis Sensitivity involved the preparation of partial maps, which are fundamental for understanding the particular situation of each component of the physical environment in the study area. Before presenting the result of the final sensitivity map, it is necessary to comment on the partial results drawn up for both the variables and the sensitivity indicators.

As already mentioned, the sensitivity assessment of the physical environment of the study area was carried out by evaluating two indicators, as recommended in the methodological roadmap developed for the Integrated Environmental Assessment of Hydroelectric Projects (ELETROBRÁS, 2007), namely: Geology and Erosion Processes (**Table 3.3.a, Volume II**).

The first map resulting from the assessment of the geology indicator drawn up for the study area was Map 2.1.a - Sensitivity of the Natural Earthquakes and Recently Moved Faults Variable.

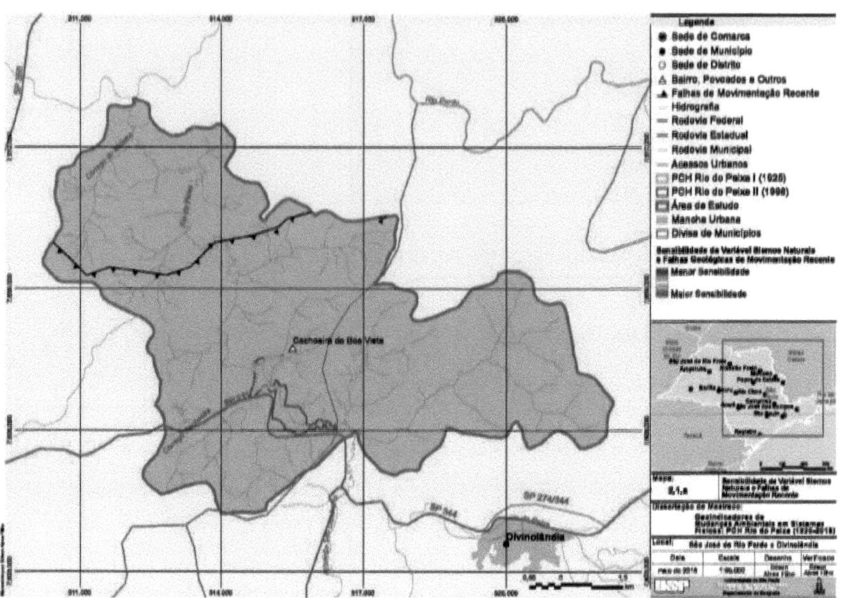

Map 2.1.a - Sensitivity of the Variable Natural Earthquakes and Recently Moved Faults in the Study Area

Map 2.1.a shows that the regions with the greatest sensitivity to natural earthquakes and recent fault movements are located along the reservoir area of the Rio do Peixe I and II SHPPs, and in downstream stretches, next to the restricted river plains of this important tributary of the Pardo River. This distribution is explained by the influence of natural earthquakes in the western part of the study area, at a distance of less than 20 km, as well as by the existence of a compressional geological fault to the north of the study area, about 2.5 km away. The influence of the crustal movements of these two elements has more pronounced effects on recent sedimentary terrains, as is the case with the river plains, which therefore explains the greater sensitivity found.

It is worth mentioning that the soils that form the substrate of the study area are stable and the existing compressional fault is inactive, posing no risk to the Rio do Peixe I and II SHP dams. However, the existence of earthquakes in regions close to the area being analysed is a factor to be taken into account when sizing the spillway structures and reinforcing the foundations made in the shoulders, symbolising an additional reinforcement for the dam's safety. Reading the Environmental Impact Study of the

The second map resulting from the assessment of the geology indicator was **Map 2.1.b - Sensitivity of the** Variable Areas Susceptible to Instabilisation of Rock Masses.

6

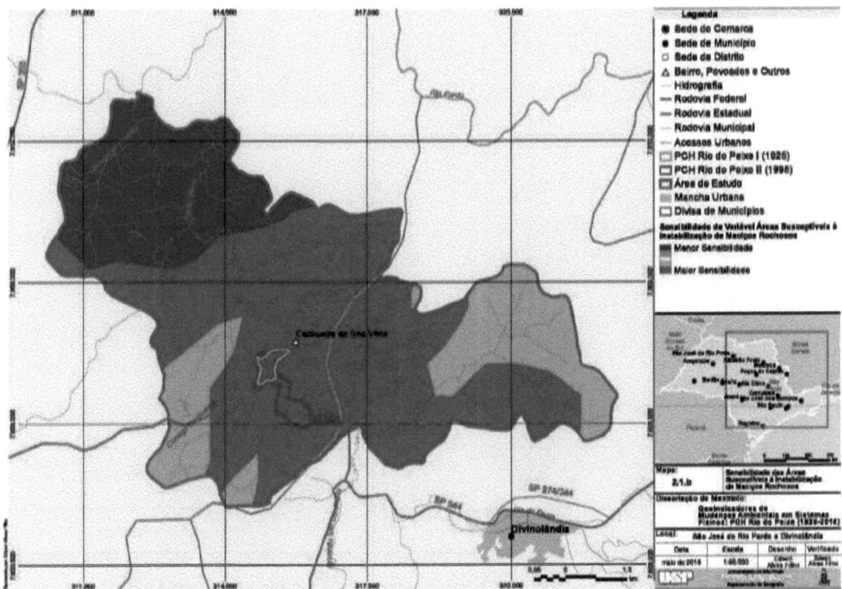

Map 2.1.b - Sensitivity of the Variable Areas Susceptible to Rock Mass Instabilisation

From Map 2.1,b it can be seen that the areas with the least sensitivity to the instabi- lisation of massifs are located in the northern part of the study area, where parag- naisses, biotite-gnaisses and schists are located, rocks with few fractures and more resistant to weathering. The areas with a medium degree of sensitivity are located to the west and southeast of the study area, where migmatites and anatexites and tonalite-gneisses are located, which are banded rocks with planes of weakness through which weathering is more easily able to act.

Finally, in the north-west, central and southern portions of the study area, there are regions with greater sensitivity to the instabilisation of massifs, where charnokites, jotunites, syenites and nori-tos, highly fissured rocks, are very susceptible to weathering.

Following on from the assessment of the Geological Sensitivity Indicator, Map 2.1.c - Sensitivity of the Mineral Resources Variable is mentioned in relation to its potential for dismantling geomorphological compartments, due to the extraction of mineral subsistences such as sand and gravel, which are very important in the context of the study area.

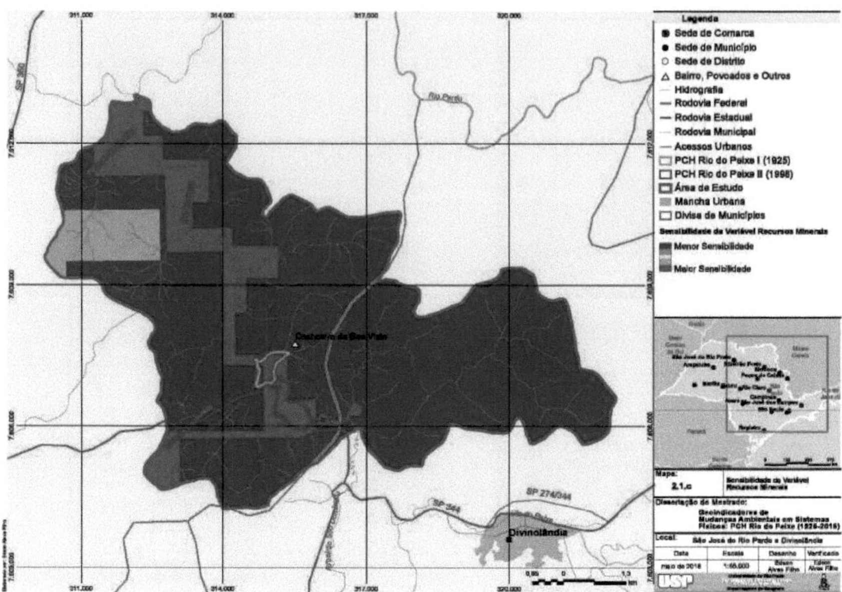

Map 2.1.c - Sensitivity of the Variable Mineral Resources

The mineral exploration areas within the study area are located along the Peixe River plain, in a north-west - south-east direction, as well as in the north-western part of it. In the first region, there are sand quarries on the River Peixe plain, an activity that produces large quantities of tailings and has the capacity to de-characterise and, in many cases, destroy this compartment of the relief, a fact that led to this variable being assigned the highest degree of sensitivity. The second exploration region, to the north-west, has granite deposits.

The exploitation of this material, especially in regions where the granite is more fractured, can trigger erosive processes such as block falls (a natural event due to the existence of fractures in the rocks, which can be enhanced by the exploitation of gravel in the region), which led to the classification of the degree of sensitivity as medium.

The fourth variable selected to assess the Geological Sensitivity Indicator is shown in **Map 2.1.d** - Sensitivity of the Hydrogeology Variable, which is considered in relation to possible impacts on the contamination of groundwater resources.

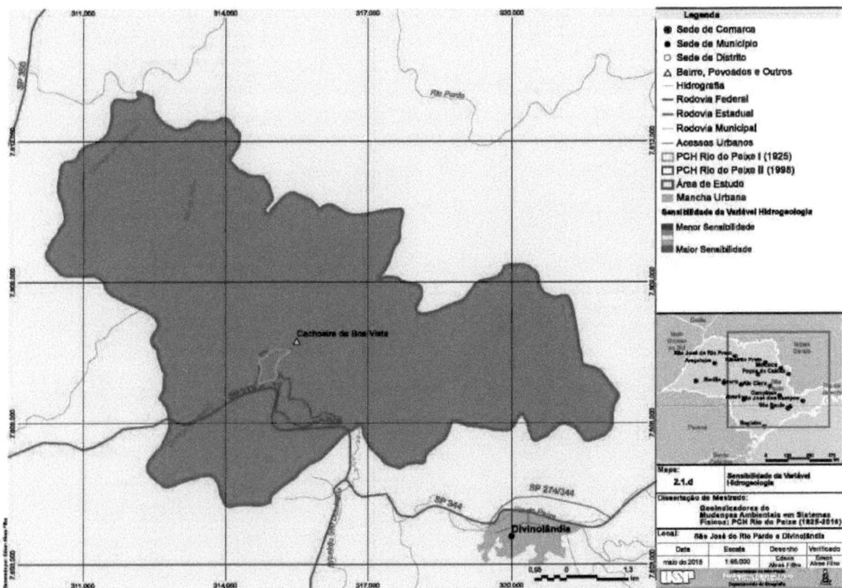

Map 2.1.d - Sensitivity of the Hydrogeology Variable

With regard to Hydrogeological Sensitivity, due to the source consulted and scale restrictions, only the Crystalline Aquifer Domain is found, which presents very localised risks of contamination, such as in areas of diaclases and fracturing of the rocks, which led to a low degree of sensitivity being assigned to the entire study area.

The fifth and final variable used to analyse the Geological Sensitivity Indicator is shown in Map 2.1.e - Sensitivity of the Geomorphology Variable. This map assigns sensitivity classes to the geomorphological compartments of the study area, taking into account the topographic position, slope, surface processes and geological substrate that make them up.

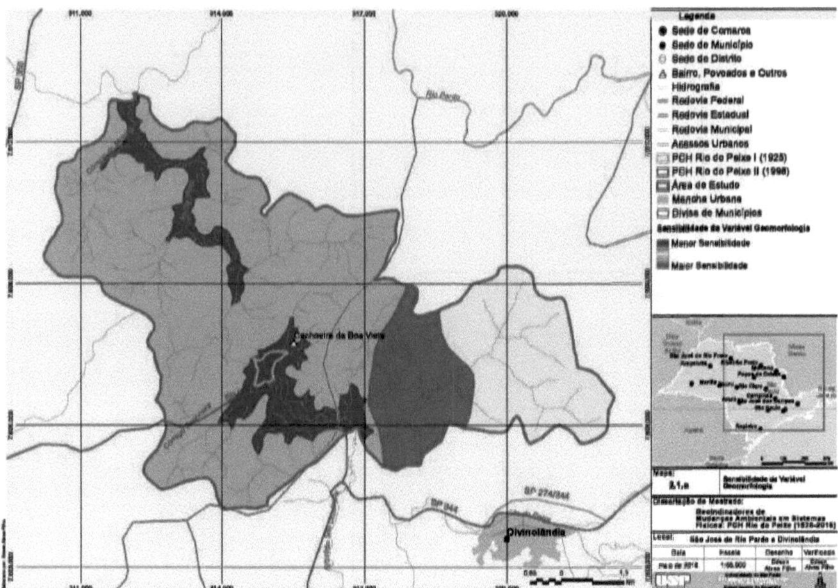

Map 2.1.e - Sensitivity of the Geomorphology Variable

With regard to the sensitivity of the Geomorphology Variable, it should be noted that the most sensitive areas are located in two strips running NW-SE, parallel to the River Peixe, where there is a succession of hills and mountains. These are precisely the areas with the highest altitudes and slopes in the study area, with a predominance of argisols and cambisols and where erosion processes are most likely to occur.

Next, the areas of hills and mountains, whose sensitivity is defined as high, appear in the N and L portions of the study area, representing sectors with high slopes, but in a lower altimetric position in relation to the hills and mountains unit.

The medium sensitivity class is represented by the units of small hills and morrotes, which are located in the NO portion of the study area, accompanying stretches of less pronounced slopes and thicker soils, such as latosols. Finally, the low and very low sensitivity classes are represented respectively by the small hills and river terraces and plains, accompanying the lower altitude stretches throughout the study area.

Based on the spatial intersection between the variables Natural Earthquakes and Recently Moved Geological Faults, Areas Susceptible to Rock Mass Instabilisation, Mineral Resources, Hydrogeology and Geomorphology, Map 2.1.f- Sensitivity of the Geology Indicator was created.

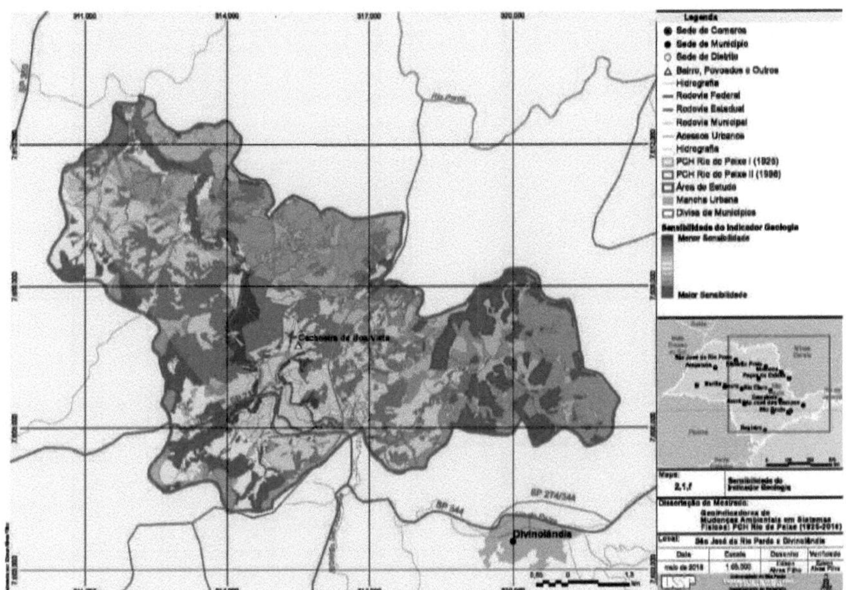

Map 2.1.f - Sensitivity of the Geology Indicator

Map 2.1.f shows that the most sensitive areas in the Geology Indicator are restricted to the river plains, located in a north-west - south-east belt, where sensitive areas are accumulated in relation to earthquakes and recently moved natural faults, with areas of great sensitivity in relation to the exploitation of mineral resources, especially sand. The first variable considered for the Erosion Processes indicator is shown in Map 2.1.g - Sensitivity of the Erodibility Variable. It should be emphasised that the degrees of sensitivity of this variable were defined on the basis of the textural characteristics and degree of development of the soils in the study area.

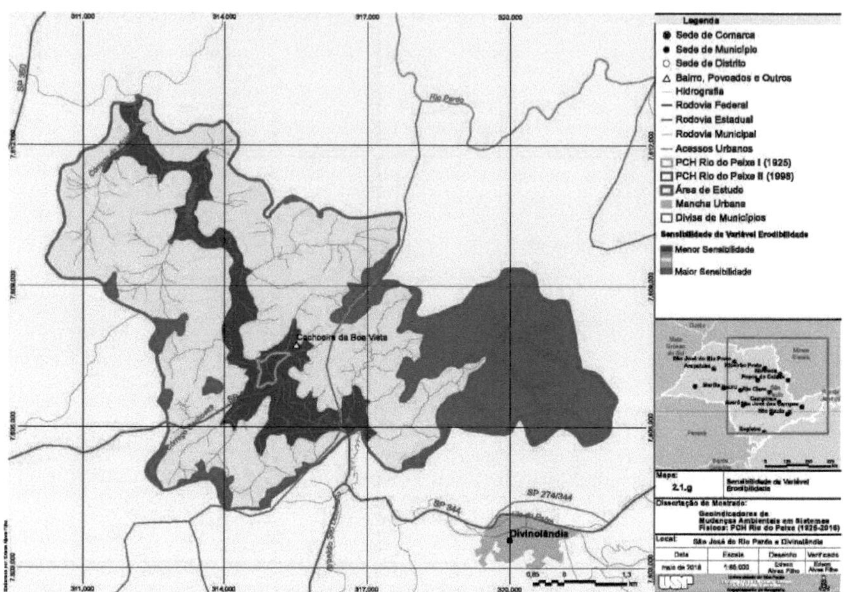

Map 2.1.g - Sensitivity of the Variable Erodibility

Map 2.1.g shows less sensitivity to erosion processes in a strip running from the north-west to the south-east of the study area, where the Fluvial Neosols are located. Due to their valley floor position, these soils do not present major restrictions to erodibility problems, which led to their classification as low erodibility.

Then there are the north-western, central and southern portions of the basin, where the Latosols are located in association with the Argissols. The high porosity and low risk of erosion of the former, due to the intermediate positions of the relief, make this type of soil with a low degree of erodibility. The second group, the Argissolos, due to the great textural discontinuity between the horizons, present more favourable conditions for the development of erosion processes. Finally, in the southeast and southwest of the study area, there are patches of Cambissolos and Neossolos Litólicos, which are poorly developed soils and, when associated with high slopes, can present conditions for the development of erosive processes.**Map 2.1.h** - Sensitivity to the Erosivity Variable, calculated from the DAEE rainfall stations in the study area and application of the Erosivity Equation, used within the Universal Soil Loss Equation, is shown below.

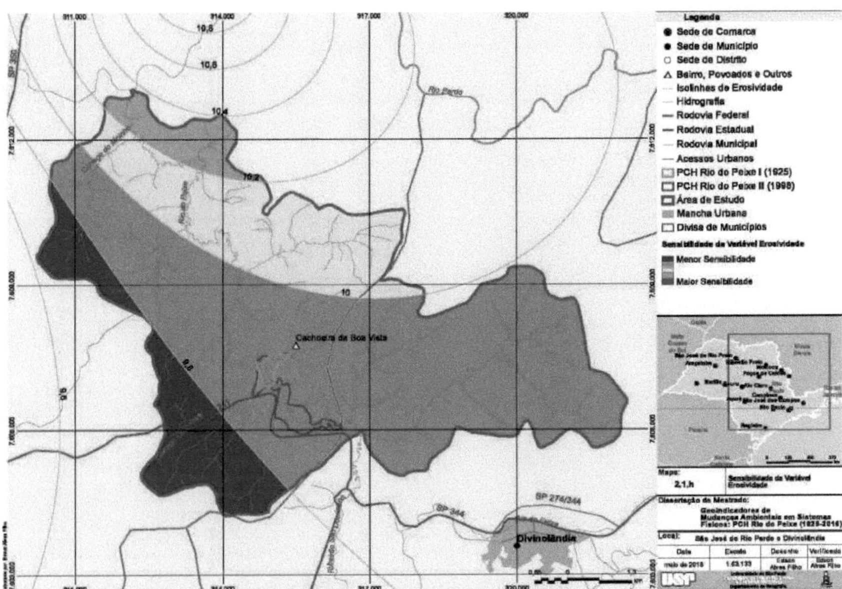

Map 2.1.h - Sensitivity of the Erosivity Variable

The highest erosivity values are located in the northern part of the study area, near the confluence of the River Peixe and the River Pardo, in a very steep area, where the calculated index is over 10.2. Next, in a north-northeast direction, near the Rio do Peixe I and II hydroelectric power stations, there is another area where rainfall is a more intense factor in erosion, with an erosivity index of 10. For the rest of the basin, the erosion caused by rainfall was classified as low, as the indices were lower than 10.

Finally, *the* last variable listed for the Erosive Processes indicator is shown in **Map 2.1.i - Sensitivity of the Variable Degree of Protection of Vegetation Cover and Land Use**. The degree of soil cover provided by the different land use classes was based on the size of the cover, with a lower degree of sensitivity for forest cover and a higher degree of sensitivity for herbaceous or anthropised cover.

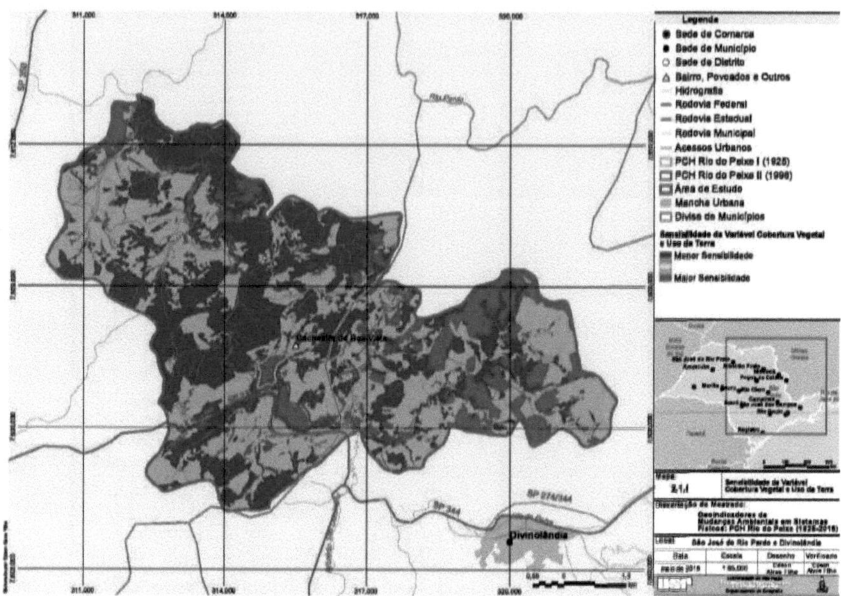

Map 2.1.i - Sensitivity of the Vegetation Cover and Land Use Variable

For the study area, the most sensitive areas in terms of the degree of protection of vegetation cover and land use are located to the north-east and south-west, along the water dividers, where small patches of exposed soil can be seen.

The less sensitive areas are spread throughout the study area, with the exception of the north-west and south-east extremes. Throughout the region of lower sensitivity, there are still important fragments of Seasonal Se- mideciduous Forest, preferentially located on dividers and stretches of high and medium slopes. The rest of the study area has a medium degree of sensitivity, represented by herbaceous and shrub cover.

The Sensitivity of the Erosive Processes Indicator is shown on **Map 2.1.j, the** result of combining maps 2.1.g and 2.1.h by spatial intersection in the ArcGis 10.2 environment. A weight of 0.25 was defined for the variables listed in the indicator, multiplying it by the partial values of the sensitivities.

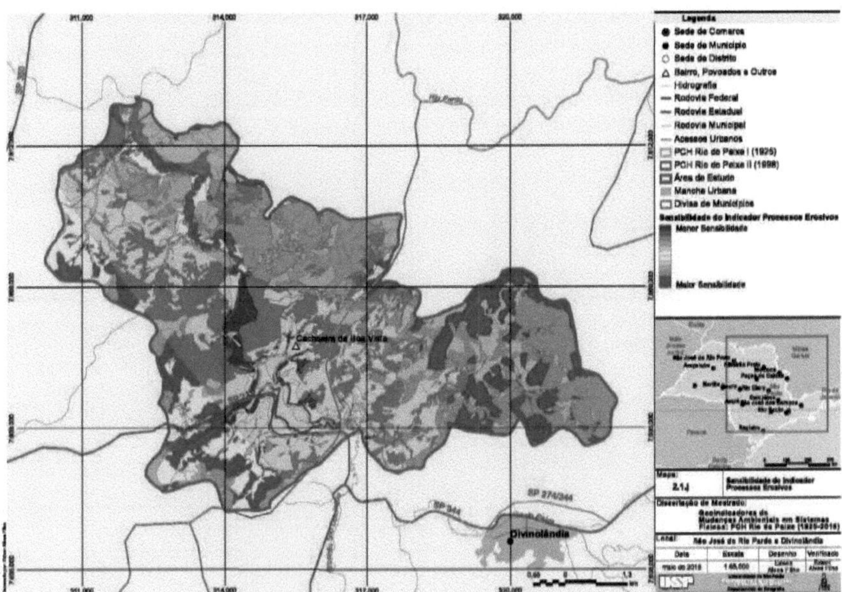

Map 2.1.j - Sensitivity of the Erosive Processes Indicator

Map 2.1.j shows that the areas with the greatest sensitivity to erosion are in the southeast of the study area, where there is a relief of elongated hills, with Argissolos in association with cambissolos. In addition, there are areas of high sensitivity to erosion in the south-west of the study area, where there are areas of exposed soil and lithic neosols.

Finally, as already mentioned, the Synthetic Sensitivity of the Physical Environment in the Study Area, shown in **Map 2.1.l,** was drawn up by spatially intersecting Maps 2.1.f and 2.1.i. The same weight was assigned to the two indicators, multiplying them by their partial sensitivities. After this procedure, a choropleth scale was applied to the final sensitivity values.

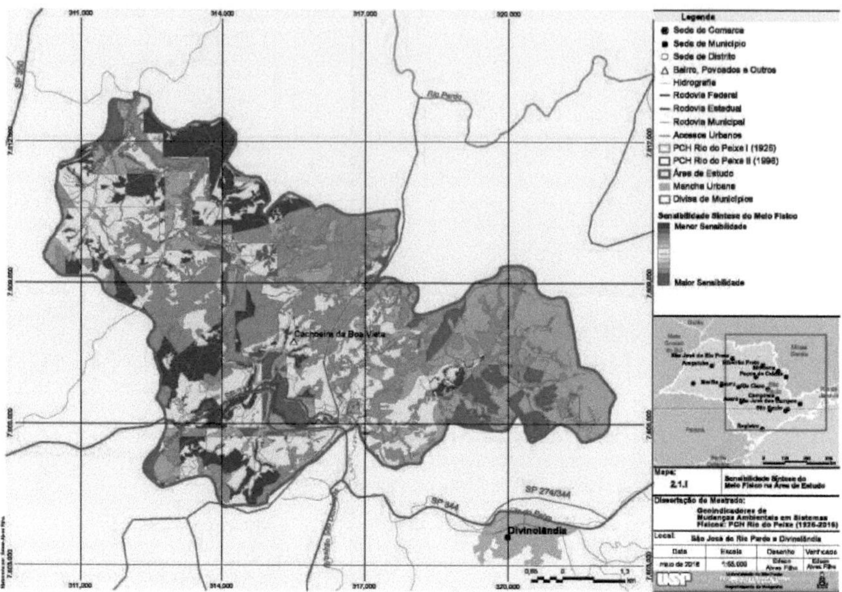

Map 2.1.1 - Synthetic Sensitivity of the Physical Environment in the Study Area

The most sensitive areas correspond to a strip running northwest-southeast from the confluence of the Peixe River with the Pardo River, heading towards the headwaters of the tributaries on the right bank of the Peixe River in the study area. This sector overlaps areas that are sensitive to earthquakes and recently moved faults, with rocks with a higher degree of fracturing, such as chamoquites, and finally, areas with sand deposits, an element that represents a risk in relation to the river plain compartment. In the south-eastern part of the study area, there are overlapping areas of hills and Argissolos in combination with cambissolos, representing a zone with a high degree of sensitivity to erosion.

The results obtained from Map 2.1.m - Negative Environmental Impacts in the Study Area are presented below. It should be noted, however, that the preparation of this map was preceded by the completion of a Qualitative and Quantitative Impacts matrix.

This matrix (**Table 2.1.a**) accounts for 53 negative impacts of the physical environment component, divided into the subcomponents Surface Water Resources (Cl.01), Groundwater Resources (C2.01), Relief and Soils (C3.01), Climate and Air Quality (C4.01) and Palaeontological Heritage (C5.01).

For the Surface Water Resources subcomponent, the impacts with the highest values were respectively:

- 1.02 - Formation of a lentic environment and alteration of the physical, chemical and biological

16

properties of surface waters (-99),

- 1.04 - Alteration of the water flow regime downstream and upstream due to the installation of the cofferdams (-81),
- 1.09 - Decrease in the flow of the watercourse downstream (-81),
- 1.16- Thermal Stratification of Reservoir Water (-99),
- 1.17- Worsening water quality downstream of the reservoir due to release

of degraded water with the emptying of the reservoir (-99),

- 1.18- Alteration of water quality due to the deposit of reservoir sediments with physico-chemical characteristics different from the original soils (-99) e;
- 1.19- Imposition of restrictions on the operation of downstream power stations by changing the

quality of the plant

water quality or the volume of sediment transported (-99).

As can be seen, the change in water quality due to the change in sedimentation dynamics with the implementation and deactivation of the reservoir ends up causing the greatest impacts on water resources as a result of the implementation of hydroelectric projects. This information can be corroborated by the water quality values found in the geo-indicators selected for the study area, as can be seen later in **Section 2.5,** which shows a worsening in water quality between the active intervention and consolidated intervention phases.

With regard to the Groundwater Resources subcomponent, the impacts with the highest values were respectively:

- 2.03 - Inducing the occurrence of processes associated with the subsurface drainage network (-90) e,
- 2.06 - Change in the Water Balance of the River Basin (-90).

Both impacts are difficult to verify due to the lack of piezo-metric data in the studies consulted or the lack of flow data from the artesian wells registered with the DAEE, which only present serial data on dates only during the project's consolidated intervention phase, thus making qualitative analyses impossible.

Table 2.1.a - Qualification and Quantification of the Negative Environmental Impacts on the Physical Environment identified for the Directly Affected Area (ADA) of the Rio do Peixe I and II SHPs (study area)

Impactful Actions	Environmental Component: Physical Environment									
	C.1.01									
	Surface Water Resources									
A2. Construction Phase	Impact	Incidence	Value	Distributivity	Value	Time of Incidence	Value	Length of stay	Value	Probability
A.2.04 - Clearing and	1.05	DI	2	LO	1	IM	2	TE	1	EC
suppression of	1.06	DI	2	LO	1	IM	2	TE	1	EC
vegetation in areas of	1.07	DI	2	LO	1	IM	2	TE	1	EC

direct intervention (permanent structures, construction sites)										
A.2.05 - Cleaning and clearing of dump areas	1.05	DI	2	LO	1	IM	2	TE	1	EC
	1.06	DI	2	LO	1	IM	2	TE	1	EC
	1.07	DI	2	LO	1	IM	2	TE	1	EC
A.2.06 - Setting up construction sites	1.06	DI	2	LO	1	IM	2	TE	1	EC
	1.07	DI	2	LO	1	IM	2	TE	1	EC
A.2.07 - Exploitation of borrow areas (clay soils)	1.06	DI	2	LO	1	IM	2	TE	1	EC
	1.07	DI	2	LO	1	IM	2	TE	1	EC
A.2.08 - Sand and gravel extraction	1.06	DI	2	LO	1	IM	2	TE	1	EC
A.2.09 - Handling of dumpsites	1.06	DI	2	LO	1	IM	2	TE	1	EC
	1.07	DI	2	LO	1	IM	2	TE	1	EC
A.2.10- Management of soil dumps and deposits of other construction materials	1.06	DI	2	LO	1	IM	2	TE	1	EC
	1.07	DI	2	LO	1	IM	2	TE	1	EC

Value	Magnitude	Synergy	Value	Reversibility	Value	Relevance	Value	Importance	Sense	Impact Value	Value of Normalised Impact
3	9	MA	3	R	1	G	4	8	-1	-72	3,6
3	9	MA	3	R	1	M	3	7	-1	-63	3,1
3	9	MA	3	R	1	M	3	7	-1	-63	3,1
3	9	MA	3	R	1	G	4	8	-1	-72	3,6
3	9	MA	3	R	1	M	3	7	-1	-63	3,1
3	9	MA	3	R	1	M	3	7	-1	-63	3,1
3	9	MA	3	R	1	M	3	7	-1	-63	3,1
3	9	MA	3	R	1	M	3	7	-1	-63	3,1
3	9	MA	3	R	1	M	3	7	-1	-63	3,1
3	9	MA	3	R	1	M	3	7	-1	-63	3,1
3	9	MA	3	R	1	M	3	7	-1	-63	3,1
3	9	MA	3	R	1	M	3	7	-1	-63	3,1
3	9	MA	3	R	1	M	3	7	-1	-63	3,1
3	9	MA	3	R	1	M	3	7	-1	-63	3,1
3	9	MA	3	R	1	M	3	7	-1	-63	3,1

A.2.11 - Construction of new accesses and improvements to existing accesses and paths	1.06	DI	2	LO	1	IM	2	TE	1	EC
	1.07	DI	2	LO	1	IM	2	TE	1	EC
A.2.12 - Implementation of a sand transhipment structure	1.06	DI	2	LO	1	IM	2	TE	1	EC
	1.07	DI	2	LO	1	IM	2	TE	1	EC
A.2.13 - Construction of a bridge to cross the River Peixe	1.06	DI	2	LO	1	IM	2	TE	1	EC
	1.07	DI	2	LO	1	IM	2	TE	1	EC
A.2.14 - Construction of internal site accesses	1.06	DI	2	LO	1	IM	2	TE	1	EC
	1.07	DI	2	LO	1	IM	2	TE	1	EC
A.2.16-Operation of the construction site	1.06	DI	2	LO	1	IM	2	TE	1	EC
A.2.17 -	1.04	DI	2	LO	1	IM	2	TE	1	EC
	1.06	DI	2	LO	1	IM	2	TE	1	EC

Implementation of upstream and downstream cofferdams and river diversion		1.07	DI	2	LO	1	IM	2	TE	1	EC
A.2.18 - Compulsory excavations		1.06	DI	2	LO	1	IM	2	TE	1	EC
		1.07	DI	2	LO	1	IM	2	TE	1	EC
A.2.20 - Waterproofing injections		1.06	DI	2	LO	1	IM	2	TE	1	EC
A.2.21 - Installation of concrete structures		1.06	DI	2	LO	1	IM	2	TE	1	EC
A.2.22 - Construction of the Escape Channel, Powerhouse and Spillway		1.06	DI	2	LO	1	IM	2	TE	1	EC
		1.07	DI	2	LO	1	IM	2	TE	1	EC
A.2.23 - Sub-surface drainage of concrete structures		1.06	DI	2	LO	1	IM	2	TE	1	EC
		1.07	DI	2	LO	1	IM	2	TE	1	EC
A.2.24 - Anchoring the escape channel		1.06	DI	2	LO	1	IM	2	TE	1	EC
		1.07	DI	2	LO	1	IM	2	TE	1	EC
3	9	MA	3	R	1	M	3	7	-1	-63	3,1
3	9	MA	3	R	1	M	3	7	-1	-63	3,1
3	9	MA	3	R	1	M	3	7	-1	-63	3,1
3	9	MA	3	R	1	M	3	7	-1	-63	3,1
3	9	MA	3	R	1	M	3	7	-1	-63	3,1
3	9	MA	3	R	1	M	3	7	-1	-63	3,1
3	9	MA	3	R	1	M	3	7	-1	-63	3,1
3	9	MA	3	R	1	M	3	7	-1	-63	3,1
3	9	MA	3	R	1	M	3	7	-1	-63	3,1
3	9	MA	3	R	1	MG	5	9	-1	-81	4,0
3	9	MA	3	R	1	M	3	7	-1	-63	3,1
3	9	MA	3	R	1	M	3	7	-1	-63	3,1
3	9	MA	3	R	1	M	3	7	-1	-63	3,1
3	9	MA	3	R	1	M	3	7	-1	-63	3,1
3	9	MA	3	R	1	M	3	7	-1	-63	3,1
3	9	MA	3	R	1	M	3	7	-1	-63	3,1
3	9	MA	3	R	1	M	3	7	-1	-63	3,1
3	9	MA	3	R	1	M	3	7	-1	-63	3,1
3	9	MA	3	R	1	M	3	7	-1	-63	3,1
3	9	MA	3	R	1	M	3	7	-1	-63	3,1
3	9	MA	3	R	1	M	3	7	-1	-63	3,1
A.2.25 - Draining, pumping and cofferdamming the area between the upstream and downstream cofferdams		1.06	DI	2	LO	1	IM	2	TE	1	EC
A.2.26 - Surface protection of dams		1.06	DI	2	LO	1	IM	2	TE	1	EC
K.221 - Sub-surface drainage of jambs		1.06	DI	2	LO	1	IM	2	TE	1	EC
A.2.28 - Electromechanical assembly		1.06	DI	2	LO	1	IM	2	TE	1	EC
		1.05	DI	2	LO	1	IM	2	TE	1	EC

Activity										
A.2.29 - Execution of complementary civil works	1.06	DI	2	LO	1	IM	2	TE	1	EC
	1.07	DI	2	LO	1	IM	2	TE	1	EC
A.2.30 - Clearing vegetation and cleaning the flood area	1.05	DI	2	LO	1	IM	2	TE	1	EC
	1.06	DI	2	LO	1	IM	2	TE	1	EC
	1.07	DI	2	LO	1	IM	2	TE	1	EC
A.2.31- Rehabilitation of local roads and accesses	1.06	DI	2	LO	1	IM	2	TE	1	EC
	1.07	DI	2	LO	1	IM	2	TE	1	EC
A.2.32 - Filling the tank	1.01	DI	2	LO	1	ME	1	PE	3	EC
	1.04	DI	2	LO	1	IM	2	TE	1	EC
	1.08	DI	2	LO	1	IM	2	TE	1	EC
	1.09	DI	2	LO	1	IM	2	TE	1	EC
	1.10	DI	2	LO	1	ME	1	PE	3	EC
	1.11	DI	2	RE	2	IM	2	PE	3	EC
	1.12	DI	2	RE	2	IM	2	PE	3	EC
	1.13	DI	2	LO	1	IM	2	PE	3	EC
	1.14	DI	2	LO	1	IM	2	PE	3	EC
	1.15	DI	2	LO	1	IM	2	TE	1	EC
	1.16	DI	2	LO	1	IM	2	PE	3	EC
	1.19	IN	1	RE	2	IM	2	PE	3	EC
A.2.33 - Pre-operational tests	1.06	DI	2	LO	1	IM	2	TE	1	EC

3	9	MA	3	R	1	M	3	7	-1	-63	3,1
3	9	MA	3	R	1	M	3	7	-1	-63	3,1
3	9	MA	3	R	1	M	3	7	-1	-63	3,1
3	9	MA	3	R	1	M	3	7	-1	-63	3,1
3	9	MA	3	R	1	G	4	8	-1	-72	3,6
3	9	MA	3	R	1	M	3	7	-1	-63	3,1
3	9	MA	3	R	1	M	3	7	-1	-63	3,1
3	9	MA	3	R	1	G	4	8	-1	-72	3,6
3	9	MA	3	R	1	M	3	7	-1	-63	3,1
3	9	MA	3	R	1	M	3	7	-1	-63	3,1
3	9	MA	3	R	1	M	3	7	-1	-63	3,1
3	9	MA	3	R	1	M	3	7	-1	-63	3,1
3	10	MA	3	R	1	G	4	8	-1	-80	4,0
3	9	MA	3	R	1	MG	5	9	-1	-81	4,1
3	9	ME	1	R	1	M	3	5	-1	-45	2,2
3	9	MA	3	R	1	MG	5	9	-1	-81	4,0
3	10	MA	3	R	1	M	3	7	-1	-70	3,5
3	12	ME	1	IR	3	M	3	7	-1	-84	4,2
3	12	ME	1	IR	3	M	3	7	-1	-84	4,2
3	11	MA	3	IR	3	M	3	9	-1	-99	5,0
3	11	ME	1	R	1	M	3	5	-1	-55	2,7
3	9	MA	3	R	1	MG	5	9	-1	-81	4,0
3	11	MA	3	IR	3	M	3	9	-1	-99	5,0
3	11	MA	3	IR	3	M	3	9	-1	-99	5,0
3	9	MA	3	R	1	M	3	7	-1	-63	3,1

Activity										
A.2.34 - Decommissioning and cleaning up construction sites	1.07	DI	2	LO	1	IM	2	TE	1	EC
A.3.01 - Operation of the Rio do Peixe SHPP	1.01	DI	2	LO	1	ME	1	PE	3	EC
	1.03	DI	2	LO	1	ME	1	PE	3	EC
	1.04	DI	2	LO	1	IM	2	TE	1	EC
	1.09	DI	2	LO	1	IM	2	TE	1	EC
	1.10	DI	2	LO	1	ME	1	PE	3	EC
	1.11	DI	2	RE	2	IM	2	PE	3	EC

Action	Impact	Incidence	Value	Distributivity	Value	Time of Incidence	Value	Length of stay	Value	Probability
	1.15	DI	2	LO	1	IM	2	TE	1	EC
	1.16	DI	2	LO	1	IM	2	PE	3	EC
	1.19	IN	1	RE	2	IM	2	PE	3	EC
A.3.02 - Routine preventive maintenance and upkeep	1.02	DI	2	LO	1	IM	2	PE	3	EC
A.3.03 - Maintenance Corrective	1.02	DI	2	LO	1	IM	2	PE	3	EC
A.4.01 - Reservoir emptying	1.17	DI	2	LO	1	IM	2	PE	3	EC
A.4.01 - Reservoir emptying	1.18	DI	2	LO	1	IM	2	PE	3	EC
A.4.01 - Reservoir emptying	1.19	IN	1	RE	2	IM	2	PE	3	EC
A.4.02 - Removal of the Dam, Spillway, Escape Channel, Power House and other civil works	1.18	DI	2	LO	1	IM	2	PE	3	EC
A.4.02 - Removal of the Dam, Spillway, Escape Channel, Power House and other civil works	1.19	IN	1	RE	2	IM	2	PE	3	EC

Impactful Actions	Environmental Component: Physical Environment									
	C.1.02									
	Groundwater Resources									
A2. Construction Phase	Impact	Incidence	Value	Distributivity	Value	Time of Incidence	Value	Length of stay	Value	Probability
A.2.04 - Clearing and suppression of vegetation in areas of direct intervention (permanent structures, construction sites)	2.03	DI	2	LO	1	ME	1	PE	3	PR

		Impact	Incidence	Value	Distributivity	Value	Time of Incidence	Value	Length of stay	Value	Probability
3	9	MA	3	R	1	M	3	7	-1	-63	3,1
3	10	MA	3	R	1	G	4	8	-1	-80	4,0
3	10	MA	3	R	1	G	4	8	-1	-80	4,0
3	9	MA	3	R	1	MG	5	9	-1	-81	4,1
3	9	MA	3	R	1	MG	5	9	-1	-81	4,1
3	10	MA	3	R	1	M	3	7	-1	-70	3,5
3	12	ME	1	IR	3	M	3	7	-1	-84	4,2
3	9	MA	3	R	1	MG	5	9	-1	-81	4,1
3	11	MA	3	IR	3	M	3	9	-1	-99	5,0
3	11	MA	3	IR	3	M	3	9	-1	-99	5,0
3	11	MA	3	IR	3	M	3	9	-1	-99	5,0
3	11	MA	3	IR	3	M	3	9	-1	-99	5,0
3	11	MA	3	IR	3	M	3	9	-1	-99	5,0
3	11	MA	3	IR	3	M	3	9	-1	-99	5,0
3	11	MA	3	IR	3	M	3	9	-1	-99	5,0
3	11	MA	3	IR	3	M	3	9	-1	-99	5,0
3	11	MA	3	IR	3	M	3	9	-1	-99	5,0

Value	Magnitude	Synergy	Value	Reversibility	Value	Relevance	Value	Importance	Sense	Impact Value	Value of Normalised Impact
2	9	MA	3	IR	3	G	4	10	-1	-90	4,5
A.2.06 - Setting up construction sites	2.03	DI	2	LO	1	ME	1	PE	3	PR	
A.2.06 - Setting up construction sites	2.07	IN	1	RE	2	IM	2	TE	1	PR	
A.2.07 - Exploitation of borrow areas (clay soils)	2.03	DI	2	LO	1	ME	1	PE	3	PR	
A.2.09 - Management of the dumpsites	2.03	DI	2	LO	1	ME	1	PE	3	PR	
A.2.10- Management	2.03	DI	2	LO	1	ME	1	PE	3	PR	

of soil dumps and deposits of other construction materials											
A.2.11 - Construction of new accesses and improvements to existing accesses and paths		2.03	DI	2	LO	1	ME	1	PE	3	PR
A.2.12 - Implementation of a sand transhipment structure		2.03	DI	2	LO	1	ME	1	PE	3	PR
A.2.13 - Construction of a bridge to cross the River do Fish		2.03	DI	2	LO	1	ME	1	PE	3	PR
A.2.16 - Operation of the construction site		2.07	IN	1	RE	2	IM	2	TE	1	PR
A.2.17 - Implementation of upstream and downstream cofferdams and river diversion		2.03	DI	2	LO	1	ME	1	PE	3	PR
A.2.18 - Compulsory excavations		2.02	DI	2	LO	1	IM	2	PE	3	EC
		2.03	DI	2	LO	1	ME	1	PE	3	PR
		2.04	DI	2	LO	1	IM	2	PE	3	EC
		2.05	DI	2	LO	1	IM	2	PE	3	EC
		2.06	IN	1	RE	2	ME	1	PE	3	EC
		2.07	IN	1	RE	2	IM	2	TE	1	PR
A.2.19 - Transport of excavated materials		2.03	DI	2	LO	1	ME	1	PE	3	PR
2	9	MA	3	IR	3	G	4	10	-1	-90	4,5
2	8	ME	1	R	1	M	3	5	-1	-40	2,0
2	9	MA	3	IR	3	G	4	10	-1	-90	4,5
2	9	MA	3	IR	3	G	4	10	-1	-90	4,5
2	9	MA	3	IR	3	G	4	10	-1	-90	4,5
2	9	MA	3	IR	3	G	4	10	-1	-90	4,5
2	9	MA	3	IR	3	G	4	10	-1	-90	4,5
2	9	MA	3	IR	3	G	4	10	-1	-90	4,5
2	8	ME	1	R	1	M	3	5	-1	-40	2,0
2	9	MA	3	IR	3	G	4	10	-1	-90	4,5
3	11	ME	1	IR	3	M	3	7	-1	-77	3,8
2	9	MA	3	IR	3	G	4	10	-1	-90	4,5
3	11	ME	1	IR	3	M	3	7	-1	-77	3,8
3	11	ME	1	IR	3	M	3	7	-1	-77	3,8
3	10	MA	3	IR	3	M	3	9	-1	-90	4,5
2	8	ME	1	R	1	M	3	5	-1	-40	2,0
2	9	MA	3	IR	3	G	4	10	-1	-90	4,5
A.2.20 - Waterproofing injections		2.02	DI	2	LO	1	IM	2	PE	3	EC
		2.04	DI	2	LO	1	IM	2	PE	3	EC
		2.05	DI	2	LO	1	IM	2	PE	3	EC
		2.06	IN	1	RE	2	ME	1	PE	3	EC
		2.07	IN	1	RE	2	IM	2	TE	1	PR
A.2.21 - Installation of concrete structures		2.03	DI	2	LO	1	ME	1	PE	3	PR
		2.07	IN	1	RE	2	IM	2	TE	1	PR
A.2.22 - Construction of the Escape		2.02	DI	2	LO	1	IM	2	PE	3	EC
		2.03	DI	2	LO	1	ME	1	PE	3	PR
		2.04	DI	2	LO	1	IM	2	PE	3	EC

Channel, Powerhouse and Spillway	2.05	DI	2	LO	1	IM	2	PE	3	EC
	2.06	IN	1	RE	2	ME	1	PE	3	EC
	2.07	IN	1	RE	2	IM	2	TE	1	PR
A.2.23 - Sub-surface drainage of concrete structures	2.02	DI	2	LO	1	IM	2	PE	3	EC
	2.04	DI	2	LO	1	IM	2	PE	3	EC
	2.05	DI	2	LO	1	IM	2	PE	3	EC
	2.06	IN	1	RE	2	ME	1	PE	3	EC
	2.07	IN	1	RE	2	IM	2	TE	1	PR
A.2.24 - Anchoring the escape channel	2.02	DI	2	LO	1	IM	2	PE	3	EC
	2.03	DI	2	LO	1	ME	1	PE	3	PR
	2.04	DI	2	LO	1	IM	2	PE	3	EC
	2.05	DI	2	LO	1	IM	2	PE	3	EC
	2.06	IN	1	RE	2	ME	1	PE	3	EC
	2.07	IN	1	RE	2	IM	2	TE	1	PR
K7..Z1 - Sub-surface drainage of jambs	2.07	IN	1	RE	2	IM	2	TE	1	PR
A.2.28 - Electromechanical assembly	2.07	IN	1	RE	2	IM	2	TE	1	PR
A.2.29 - Execution of complementary civil works	2.03	DI	2	LO	1	ME	1	PE	3	PR
	2.07	IN	1	RE	2	IM	2	TE	1	PR

3	11	ME	1	IR	3	M	3	7	-1	-77	3,8
3	11	ME	1	IR	3	M	3	7	-1	-77	3,8
3	11	ME	1	IR	3	M	3	7	-1	-77	3,8
3	10	MA	3	IR	3	M	3	9	-1	-90	4,5
2	8	ME	1	R	1	M	3	5	-1	-40	2,0
2	9	MA	3	IR	3	G	4	10	-1	-90	4,5
2	8	ME	1	R	1	M	3	5	-1	-40	2,0
3	11	ME	1	IR	3	M	3	7	-1	-77	3,8
2	9	MA	3	IR	3	G	4	10	-1	-90	4,5
3	11	ME	1	IR	3	M	3	7	-1	-77	3,8
3	11	ME	1	IR	3	M	3	7	-1	-77	3,8
3	10	MA	3	IR	3	M	3	9	-1	-90	4,5
2	8	ME	1	R	1	M	3	5	-1	-40	2,0
3	11	ME	1	IR	3	M	3	7	-1	-77	3,8
3	11	ME	1	IR	3	M	3	7	-1	-77	3,8
3	11	ME	1	IR	3	M	3	7	-1	-77	3,8
3	10	MA	3	IR	3	M	3	9	-1	-90	4,5
2	8	ME	1	R	1	M	3	5	-1	-40	2,0
3	11	ME	1	IR	3	M	3	7	-1	-77	3,8
2	9	MA	3	IR	3	G	4	10	-1	-90	4,5
3	11	ME	1	IR	3	M	3	7	-1	-77	3,8
3	11	ME	1	IR	3	M	3	7	-1	-77	3,8
3	10	MA	3	IR	3	M	3	9	-1	-90	4,5
2	8	ME	1	R	1	M	3	5	-1	-40	2,0
2	8	ME	1	R	1	M	3	5	-1	-40	2,0
2	8	ME	1	R	1	M	3	5	-1	-40	2,0
2	9	MA	3	IR	3	G	4	10	-1	-90	4,5
2	8	ME	1	R	1	M	3	5	-1	-40	2,0

A.2.30 - Clearing vegetation and cleaning the flood area	2.03	DI	2	LO	1	ME	1	PE	3	PR
	2.07	IN	1	RE	2	IM	2	TE	1	PR
A.2.31- Rehabilitation of local roads and accesses	2.03	DI	2	LO	1	ME	1	PE	3	PR
A.2.32 - Filling the tank	2.01	DI	2	LO	1	IM	2	PE	3	EC

Activity										
A.2.34 - Decommissioning and cleaning up construction sites	2.03	DI	2	LO	1	ME	1	PE	3	PR
	2.07	IN	1	RE	2	IM	2	TE	1	PR
A.2.04 - Clearing and suppression of vegetation in areas of direct intervention (permanent structures, construction sites)	2.03	DI	2	LO	1	ME	1	PE	3	PR
A.2.06 - Setting up construction sites	2.03	DI	2	LO	1	ME	1	PE	3	PR
	2.07	IN	1	RE	2	IM	2	TE	1	PR
A.2.07 - Exploitation of borrow areas (clay soils)	2.03	DI	2	LO	1	ME	1	PE	3	PR
A.2.09 - Handling of dumpsites	2.03	DI	2	LO	1	ME	1	PE	3	PR
A.2.10- Management of soil dumps and deposits of other construction materials	2.03	DI	2	LO	1	ME	1	PE	3	PR
A.2.11 - Construction of new accesses and improvements to existing accesses and paths	2.03	DI	2	LO	1	ME	1	PE	3	PR
A.2.12 - Implementation of a sand transhipment structure	2.03	DI	2	LO	1	ME	1	PE	3	PR
A.2.13 - Construction of a bridge to cross the River Peixe	2.03	DI	2	LO	1	ME	1	PE	3	PR

2	9	MA	3	IR	3	G	4	10	-1	-90	4,5
2	8	ME	1	R	1	M	3	5	-1	-40	2,0
2	9	MA	3	IR	3	G	4	10	-1	-90	4,5
3	11	ME	1	IR	3	M	3	7	-1	-77	3,8
2	9	MA	3	IR	3	G	4	10	-1	-90	4,5
2	8	ME	1	R	1	M	3	5	-1	-40	2,0
2	9	MA	3	IR	3	G	4	10	-1	-90	4,5
2	9	MA	3	IR	3	G	4	10	-1	-90	4,5
2	8	ME	1	R	1	M	3	5	-1	-40	2,0
2	9	MA	3	IR	3	G	4	10	-1	-90	4,5
2	9	MA	3	IR	3	G	4	10	-1	-90	4,5
2	9	MA	3	IR	3	G	4	10	-1	-90	4,5
2	9	MA	3	IR	3	G	4	10	-1	-90	4,5
2	9	MA	3	IR	3	G	4	10	-1	-90	4,5
2	9	MA	3	IR	3	G	4	10	-1	-90	4,5

Activity										
A.2.16-Operation of the construction site	2.07	IN	1	RE	2	IM	2	TE	1	PR
A.2.17 - Implementation of upstream and downstream cofferdams and river diversion	2.03	DI	2	LO	1	ME	1	PE	3	PR
A.2.18 - Compulsory	2.02	DI	2	LO	1	IM	2	PE	3	EC
	2.03	DI	2	LO	1	ME	1	PE	3	PR

excavations	2.04	DI	2	LO	1	IM	2	PE	3	EC	
	2.05	DI	2	LO	1	IM	2	PE	3	EC	
	2.06	IN	1	RE	2	ME	1	PE	3	EC	
	2.07	IN	1	RE	2	IM	2	TE	1	PR	
A.2.19 - Transport of excavated materials	2.03	DI	2	LO	1	ME	1	PE	3	PR	
A.2.20 - Waterproofing injections	2.02	DI	2	LO	1	IM	2	PE	3	EC	
	2.04	DI	2	LO	1	IM	2	PE	3	EC	
	2.05	DI	2	LO	1	IM	2	PE	3	EC	
	2.06	IN	1	RE	2	ME	1	PE	3	EC	
	2.07	IN	1	RE	2	IM	2	TE	1	PR	
A.2.21 - Installation of concrete structures	2.03	DI	2	LO	1	ME	1	PE	3	PR	
	2.07	IN	1	RE	2	IM	2	TE	1	PR	
A.2.22 - Construction of the Escape Channel, Powerhouse and Spillway	2.02	DI	2	LO	1	IM	2	PE	3	EC	
	2.03	DI	2	LO	1	ME	1	PE	3	PR	
	2.04	DI	2	LO	1	IM	2	PE	3	EC	
	2.05	DI	2	LO	1	IM	2	PE	3	EC	
	2.06	IN	1	RE	2	ME	1	PE	3	EC	
	2.07	IN	1	RE	2	IM	2	TE	1	PR	
A.2.23 - Sub-surface drainage of concrete structures	2.02	DI	2	LO	1	IM	2	PE	3	EC	
	2.04	DI	2	LO	1	IM	2	PE	3	EC	
	2.05	DI	2	LO	1	IM	2	PE	3	EC	
	2.06	IN	1	RE	2	ME	1	PE	3	EC	
	2.07	IN	1	RE	2	IM	2	TE	1	PR	
2	8	ME	1	R	1	M	3	5	-1	-40	2,0
2	9	MA	3	IR	3	G	4	10	-1	-90	4,5
3	11	ME	1	IR	3	M	3	7	-1	-77	3,8
2	9	MA	3	IR	3	G	4	10	-1	-90	4,5
3	11	ME	1	IR	3	M	3	7	-1	-77	3,8
3	11	ME	1	IR	3	M	3	7	-1	-77	3,8
3	10	MA	3	IR	3	M	3	9	-1	-90	4,5
2	8	ME	1	R	1	M	3	5	-1	-40	2,0
2	9	MA	3	IR	3	G	4	10	-1	-90	4,5
3	11	ME	1	IR	3	M	3	7	-1	-77	3,8
3	11	ME	1	IR	3	M	3	7	-1	-77	3,8
3	11	ME	1	IR	3	M	3	7	-1	-77	3,8
3	10	MA	3	IR	3	M	3	9	-1	-90	4,5
2	8	ME	1	R	1	M	3	5	-1	-40	2,0
2	9	MA	3	IR	3	G	4	10	-1	-90	4,5
2	8	ME	1	R	1	M	3	5	-1	-40	2,0
3	11	ME	1	IR	3	M	3	7	-1	-77	3,8
2	9	MA	3	IR	3	G	4	10	-1	-90	4,5
3	11	ME	1	IR	3	M	3	7	-1	-77	3,8
3	11	ME	1	IR	3	M	3	7	-1	-77	3,8
3	10	MA	3	IR	3	M	3	9	-1	-90	4,5
2	8	ME	1	R	1	M	3	5	-1	-40	2,0
3	11	ME	1	IR	3	M	3	7	-1	-77	3,8
3	11	ME	1	IR	3	M	3	7	-1	-77	3,8
3	11	ME	1	IR	3	M	3	7	-1	-77	3,8
3	10	MA	3	IR	3	M	3	9	-1	-90	4,5
2	8	ME	1	R	1	M	3	5	-1	-40	2,0
A.2.24 - Anchoring the escape channel	2.02	DI	2	LO	1	IM	2	PE	3	EC	
	2.03	DI	2	LO	1	ME	1	PE	3	PR	
	2.04	DI	2	LO	1	IM	2	PE	3	EC	
A.2.27 - Sub-surface shoulder drainage	2.07	IN	1	RE	2	IM	2	TE	1	PR	
A.2.28 - Electromechanical assembly	2.07	IN	1	RE	2	IM	2	TE	1	PR	
	2.03	DI	2	LO	1	ME	1	PE	3	PR	

Impactful Actions	Impact	Incidence	Value	Distributivity	Value	Time of Incidence	Value	Length of stay	Value	Probability
A.2.29 - Execution of complementary civil works	2.07	IN	1	RE	2	IM	2	TE	1	PR
A.2.30 - Clearing vegetation and cleaning the flood area	2.03	DI	2	LO	1	ME	1	PE	3	PR
	2.07	IN	1	RE	2	IM	2	TE	1	PR
A.2.31- Rehabilitation of local roads and accesses	2.03	DI	2	LO	1	ME	1	PE	3	PR
A.2.32 - Filling the tank	2.01	DI	2	LO	1	IM	2	PE	3	EC
A.2.34 - Decommissioning and cleaning up construction sites	2.03	DI	2	LO	1	ME	1	PE	3	PR
	2.07	IN	1	RE	2	IM	2	TE	1	PR

Impactful Actions	Environmental Component: Physical Environment									
	C.1.03									
	Relief and Soils									
A2. Construction Phase	Impact	Incidence	Value	Distributivity	Value	Time of Incidence	Value	Length of stay	Value	Probability
A.2.02 - Land Acquisition (negotiations and indemnities)	3.03	DI	2	LO	1	IM	2	PE	3	EC
	3.23	DI	2	LO	1	IM	2	PE	3	EC
A.2.04 - Clearing and suppression of vegetation in areas of direct intervention (permanent structures, construction sites)	3.01	DI	2	LO	1	IM	2	TE	1	PR
	3.02	IN	1	LO	1	ME	1	TE	1	EC
	3.04	DI	2	LO	1	IM	2	TE	1	EC
	3.05	DI	2	LO	1	ME	1	TE	1	PR
	3.08	DI	2	LO	1	IM	2	TE	1	EC
	3.21	DI	2	LO	1	IM	2	TE	1	EC

Value	Magnitude	Synergy	Value	Reversibility	Value	Relevance	Value	Importance	Sense	Impact Value	Value of Normalised Impact
3	11	ME	1	IR	3	M	3	7	-1	-77	3,8
2	9	MA	3	IR	3	G	4	10	-1	-90	4,5
3	11	ME	1	IR	3	M	3	7	-1	-77	3,8
2	8	ME	1	R	1	M	3	5	-1	-40	2,0
2	8	ME	1	R	1	M	3	5	-1	-40	2,0
2	9	MA	3	IR	3	G	4	10	-1	-90	4,5
2	8	ME	1	R	1	M	3	5	-1	-40	2,0
2	9	MA	3	IR	3	G	4	10	-1	-90	4,5
2	8	ME	1	R	1	M	3	5	-1	-40	2,0
2	9	MA	3	IR	3	G	4	10	-1	-90	4,5
3	11	ME	1	IR	3	M	3	7	-1	-77	3,8
2	9	MA	3	IR	3	G	4	10	-1	-90	4,5
2	8	ME	1	R	1	M	3	5	-1	-40	2,0

Value	Magnitude	Synergy	Value	Reversibility	Value	Relevance	Value	Importance	Sense	Impact Value	Value of Normalised Impact
3	11	ME	1	IR	3	M	3	7	-1	-77	3,8
3	11	ME	1	IR	3	M	3	7	-1	-77	3,8
2	8	ME	1	R	1	M	3	5	-1	-40	2,0
3	7	ME	1	R	1	P	2	4	-1	-28	1,4
3	9	MA	3	R	1	M	3	7	-1	-63	3,1
2	7	MA	3	R	1	M	3	7	-1	-49	2,4
3	9	ME	1	R	1	M	3	5	-1	-45	2,2
3	9	MA	3	R	1	M	3	7	-1	-63	3,1
	3.01	DI	2	LO	1	IM	2	TE	1	PR

Activity	Code	Type		LO		IM/ME		TE/PE		EC/PR
A.2.05 - Cleaning and clearing of dump areas	3.02	IN	1	LO	1	ME	1	TE	1	EC
	3.04	DI	2	LO	1	IM	2	TE	1	EC
	3.08	DI	2	LO	1	IM	2	TE	1	EC
	3.21	DI	2	LO	1	IM	2	TE	1	EC
A.2.06 - Setting up construction sites	3.01	DI	2	LO	1	IM	2	TE	1	PR
	3.02	IN	1	LO	1	ME	1	TE	1	EC
	3.03	DI	2	LO	1	IM	2	PE	3	EC
	3.04	DI	2	LO	1	IM	2	TE	1	EC
	3.09	DI	2	LO	1	IM	2	TE	1	PR
	3.23	DI	2	LO	1	IM	2	PE	3	EC
A.2.07 - Exploitation of borrow areas (clay soils)	3.01	DI	2	LO	1	IM	2	TE	1	PR
	3.02	IN	1	LO	1	ME	1	TE	1	EC
	3.04	DI	2	LO	1	IM	2	TE	1	EC
	3.09	DI	2	LO	1	IM	2	TE	1	PR
	3.21	DI	2	LO	1	IM	2	TE	1	EC
A.2.08 - Sand and gravel extraction	3.09	DI	2	LO	1	IM	2	TE	1	PR
	3.21	DI	2	LO	1	IM	2	TE	1	EC
A.2.09 - Management of the dumpsites	3.01	DI	2	LO	1	IM	2	TE	1	PR
	3.04	DI	2	LO	1	IM	2	TE	1	EC
	3.09	DI	2	LO	1	IM	2	TE	1	PR
	3.21	DI	2	LO	1	IM	2	TE	1	EC
A.2.10-Management of soil pits and deposits of other construction materials	3.01	DI	2	LO	1	IM	2	TE	1	PR
	3.04	DI	2	LO	1	IM	2	TE	1	EC
	3.09	DI	2	LO	1	IM	2	TE	1	PR
	3.21	DI	2	LO	1	IM	2	TE	1	EC

2	8	ME	1	R	1	M	3	5	-1	-40	2,0
3	7	ME	1	R	1	P	2	4	-1	-28	1,4
3	9	MA	3	R	1	M	3	7	-1	-63	3,1
3	9	ME	1	R	1	M	3	5	-1	-45	2,2
3	9	MA	3	R	1	M	3	7	-1	-63	3,1
2	8	ME	1	R	1	M	3	5	-1	-40	2,0
3	7	ME	1	R	1	P	2	4	-1	-28	1,4
3	11	ME	1	IR	3	M	3	7	-1	-77	3,8
3	9	MA	3	R	1	M	3	7	-1	-63	3,1
2	8	ME	1	IR	3	G	4	8	-1	-64	3,2
3	11	ME	1	IR	3	M	3	7	-1	-77	3,8
2	8	ME	1	R	1	M	3	5	-1	-40	2,0
3	7	ME	1	R	1	P	2	4	-1	-28	1,4
3	9	MA	3	R	1	M	3	7	-1	-63	3,1
2	8	ME	1	IR	3	G	4	8	-1	-64	3,2
3	9	MA	3	R	1	M	3	7	-1	-63	3,1
2	8	ME	1	IR	3	G	4	8	-1	-64	3,2
3	9	MA	3	R	1	M	3	7	-1	-63	3,1
2	8	ME	1	R	1	M	3	5	-1	-40	2,0
3	9	MA	3	R	1	M	3	7	-1	-63	3,1
2	8	ME	1	IR	3	G	4	8	-1	-64	3,2
3	9	MA	3	R	1	M	3	7	-1	-63	3,1
2	8	ME	1	R	1	M	3	5	-1	-40	2,0
3	9	MA	3	R	1	M	3	7	-1	-63	3,1
2	8	ME	1	IR	3	G	4	8	-1	-64	3,2
3	9	MA	3	R	1	M	3	7	-1	-63	3,1

Activity	Code	Type		LO		IM/ME		TE/PE		EC/PR
A.2.11 - Construction of new accesses and improvements to existing accesses and paths	3.01	DI	2	LO	1	IM	2	TE	1	PR
	3.02	IN	1	LO	1	ME	1	TE	1	EC
	3.03	DI	2	LO	1	IM	2	PE	3	EC
	3.04	DI	2	LO	1	IM	2	TE	1	EC
	3.09	DI	2	LO	1	IM	2	TE	1	PR
	3.21	DI	2	LO	1	IM	2	TE	1	EC
	3.23	DI	2	LO	1	IM	2	PE	3	EC
A.2.12 -	3.01	DI	2	LO	1	IM	2	TE	1	PR
	3.02	IN	1	LO	1	ME	1	TE	1	EC

Implementation of a sand transhipment structure	3.04	DI	2	LO	1	IM	2	TE	1	EC
A.2.13 - Construction of a bridge to cross the River Peixe	3.01	DI	2	LO	1	IM	2	TE	1	PR
	3.02	IN	1	LO	1	ME	1	TE	1	EC
	3.04	DI	2	LO	1	IM	2	TE	1	EC
	3.07	DI	2	LO	1	ME	1	TE	1	PP
	3.09	DI	2	LO	1	IM	2	TE	1	PR
	3.20	IN	1	LO	1	ME	1	TE	1	PR
	3.21	DI	2	LO	1	IM	2	TE	1	EC
A.2.14 - Construction of internal site accesses	3.01	DI	2	LO	1	IM	2	TE	1	PR
	3.02	IN	1	LO	1	ME	1	TE	1	EC
	3.04	DI	2	LO	1	IM	2	TE	1	EC
	3.09	DI	2	LO	1	IM	2	TE	1	PR
	3.21	DI	2	LO	1	IM	2	TE	1	EC
A.2.16 - Operation of the construction site	3.02	IN	1	LO	1	ME	1	TE	1	EC
A.2.17 - Implementation of upstream and downstream cofferdams and river diversion	3.01	DI	2	LO	1	IM	2	TE	1	PR
	3.02	IN	1	LO	1	ME	1	TE	1	EC
	3.04	DI	2	LO	1	IM	2	TE	1	EC
	3.07	DI	2	LO	1	ME	1	TE	1	PP
	3.09	DI	2	LO	1	IM	2	TE	1	PR
	3.21	DI	2	LO	1	IM	2	TE	1	EC

2	8	ME	1	R	1	M	3	5	-1	-40	2,0
3	7	ME	1	R	1	P	2	4	-1	-28	1,4
3	11	ME	1	IR	3	M	3	7	-1	-77	3,8
3	9	MA	3	R	1	M	3	7	-1	-63	3,1
2	8	ME	1	IR	3	G	4	8	-1	-64	3,2
3	9	MA	3	R	1	M	3	7	-1	-63	3,1
3	11	ME	1	IR	3	M	3	7	-1	-77	3,8
2	8	ME	1	R	1	M	3	5	-1	-40	2,0
3	7	ME	1	R	1	P	2	4	-1	-28	1,4
3	9	MA	3	R	1	M	3	7	-1	-63	3,1
2	8	ME	1	R	1	M	3	5	-1	-40	2,0
3	7	ME	1	R	1	P	2	4	-1	-28	1,4
3	9	MA	3	R	1	M	3	7	-1	-63	3,1
1	6	ME	1	R	1	MG	5	7	-1	-42	2,1
2	8	ME	1	IR	3	G	4	8	-1	-64	3,2
2	6	ME	1	R	1	MG	5	7	-1	-42	2,1
3	9	MA	3	R	1	M	3	7	-1	-63	3,1
2	8	ME	1	R	1	M	3	5	-1	-40	2,0
3	7	ME	1	R	1	P	2	4	-1	-28	1,4
3	9	MA	3	R	1	M	3	7	-1	-63	3,1
2	8	ME	1	IR	3	G	4	8	-1	-64	3,2
3	9	MA	3	R	1	M	3	7	-1	-63	3,1
3	7	ME	1	R	1	P	2	4	-1	-28	1,4
2	8	ME	1	R	1	M	3	5	-1	-40	2,0
3	7	ME	1	R	1	P	2	4	-1	-28	1,4
3	9	MA	3	R	1	M	3	7	-1	-63	3,1
1	6	ME	1	R	1	MG	5	7	-1	-42	2,1
2	8	ME	1	IR	3	G	4	8	-1	-64	3,2
3	9	MA	3	R	1	M	3	7	-1	-63	3,1

A.2.18 - Compulsory excavations	3.01	DI	2	LO	1	IM	2	TE	1	PR
	3.02	IN	1	LO	1	ME	1	TE	1	EC
	3.04	DI	2	LO	1	IM	2	TE	1	EC
	3.06	DI	2	LO	1	ME	1	TE	1	PR
	3.07	DI	2	LO	1	ME	1	TE	1	PP
	3.09	DI	2	LO	1	IM	2	TE	1	PR
	3.21	DI	2	LO	1	IM	2	TE	1	EC

A.2.20 - Waterproofing injections	3.02	IN	1	LO	1	ME	1	TE	1	EC		
A.2.21 - Installation of concrete structures	3.01	DI	2	LO	1	IM	2	TE	1	PR		
	3.02	IN	1	LO	1	ME	1	TE	1	EC		
	3.04	DI	2	LO	1	IM	2	TE	1	EC		
	3.06	DI	2	LO	1	ME	1	TE	1	PR		
	3.07	DI	2	LO	1	ME	1	TE	1	PP		
	3.09	DI	2	LO	1	IM	2	TE	1	PR		
	3.17	DI	2	LO	1	ME	1	TE	1	PR		
	3.18	DI	2	LO	1	IM	2	TE	1	PR		
	3.20	IN	1	LO	1	ME	1	TE	1	PR		
	3.21	DI	2	LO	1	IM	2	TE	1	EC		
A.2.22 - Construction of the Escape Channel, Powerhouse and Spillway	3.01	DI	2	LO	1	IM	2	TE	1	PR		
	3.02	IN	1	LO	1	ME	1	TE	1	EC		
	3.04	DI	2	LO	1	IM	2	TE	1	EC		
	3.06	DI	2	LO	1	ME	1	TE	1	PR		
	3.07	DI	2	LO	1	ME	1	TE	1	PP		
	3.09	DI	2	LO	1	IM	2	TE	1	PR		
	3.21	DI	2	LO	1	IM	2	TE	1	EC		
A.2.23 - Subsurface drainage of concrete structures	3.02	IN	1	LO	1	ME	1	TE	1	EC		
	3.06	DI	2	LO	1	ME	1	TE	1	PR		
	3.17	DI	2	LO	1	ME	1	TE	1	PR		
	3.18	DI	2	LO	1	IM	2	TE	1	PR		
	2	8	ME	1	R	1	M	3	5	-1	-40	2,0
	3	7	ME	1	R	1	P	2	4	-1	-28	1,4
	3	9	MA	3	R	1	M	3	7	-1	-63	3,1
	2	7	MA	3	R	1	MG	5	9	-1	-63	3,1
	1	6	ME	1	R	1	MG	5	7	-1	-42	2,1
	2	8	ME	1	IR	3	G	4	8	-1	-64	3,2
	3	9	MA	3	R	1	M	3	7	-1	-63	3,1
	3	7	ME	1	R	1	P	2	4	-1	-28	1,4
	2	8	ME	1	R	1	M	3	5	-1	-40	2,0
	3	7	ME	1	R	1	P	2	4	-1	-28	1,4
	3	9	MA	3	R	1	M	3	7	-1	-63	3,1
	2	7	MA	3	R	1	MG	5	9	-1	-63	3,1
	1	6	ME	1	R	1	MG	5	7	-1	-42	2,1
	2	8	ME	1	IR	3	G	4	8	-1	-64	3,2
	2	7	ME	1	R	1	MG	5	7	-1	-49	2,4
	1	7	MA	3	R	1	M	3	7	-1	-49	2,4
	2	6	ME	1	R	1	MG	5	7	-1	-42	2,1
	3	9	MA	3	R	1	M	3	7	-1	-63	3,1
	2	8	ME	1	R	1	M	3	5	-1	-40	2,0
	3	7	ME	1	R	1	P	2	4	-1	-28	1,4
	3	9	MA	3	R	1	M	3	7	-1	-63	3,1
	2	7	MA	3	R	1	MG	5	9	-1	-63	3,1
	1	6	ME	1	R	1	MG	5	7	-1	-42	2,1
	2	8	ME	1	IR	3	G	4	8	-1	-64	3,2
	3	9	MA	3	R	1	M	3	7	-1	-63	3,1
	3	7	ME	1	R	1	P	2	4	-1	-28	1,4
	2	7	MA	3	R	1	MG	5	9	-1	-63	3,1
	2	7	ME	1	R	1	MG	5	7	-1	-49	2,4
	1	7	MA	3	R	1	M	3	7	-1	-49	2,4
A.2.24 - Anchoring the escape channel	3.01	DI	2	LO	1	IM	2	TE	1	PR		
	3.02	IN	1	LO	1	ME	1	TE	1	EC		
	3.04	DI	2	LO	1	IM	2	TE	1	EC		
A.2.25 - Draining, pumping and cofferdamming the area between the	3.02	IN	1	LO	1	ME	1	TE	1	EC		

upstream and downstream cofferdams										
A.2.26 - Surface protection of dams	3.02	IN	1	LO	1	ME	1	TE	1	EC
	3.17	DI	2	LO	1	ME	1	TE	1	PR
A.2.27 - Sub-surface shoulder drainage	3.02	IN	1	LO	1	ME	1	TE	1	EC
	3.06	DI	2	LO	1	ME	1	TE	1	PR
A.2.28 - Electromechanical assembly	3.02	IN	1	LO	1	ME	1	TE	1	EC
A.2.29 - Execution of complementary civil works	3.01	DI	2	LO	1	IM	2	TE	1	PR
	3.02	IN	1	LO	1	ME	1	TE	1	EC
	3.04	DI	2	LO	1	IM	2	TE	1	EC
	3.07	DI	2	LO	1	ME	1	TE	1	PP
	3.08	DI	2	LO	1	IM	2	TE	1	EC
	3.09	DI	2	LO	1	IM	2	TE	1	PR
	3.20	IN	1	LO	1	ME	1	TE	1	PR
	3.21	DI	2	LO	1	IM	2	TE	1	EC
A.2.30 - Clearing vegetation and cleaning the flood area	3.01	DI	2	LO	1	IM	2	TE	1	PR
	3.02	IN	1	LO	1	ME	1	TE	1	EC
	3.04	DI	2	LO	1	IM	2	TE	1	EC
	3.05	DI	2	LO	1	ME	1	TE	1	PR
	3.08	DI	2	LO	1	IM	2	TE	1	EC
A.2.31- Rehabilitation of local roads and accesses	3.01	DI	2	LO	1	IM	2	TE	1	PR
	3.04	DI	2	LO	1	IM	2	TE	1	EC
	3.09	DI	2	LO	1	IM	2	TE	1	PR
	3.21	DI	2	LO	1	IM	2	TE	1	EC

2	8	ME	1	R	1	M	3	5	-1	-40	2,0
3	7	ME	1	R	1	P	2	4	-1	-28	1,4
3	9	MA	3	R	1	M	3	7	-1	-63	3,1
3	7	ME	1	R	1	P	2	4	-1	-28	1,4
3	7	ME	1	R	1	P	2	4	-1	-28	1,4
2	7	ME	1	R	1	MG	5	7	-1	-49	2,4
3	7	ME	1	R	1	P	2	4	-1	-28	1,4
2	7	MA	3	R	1	MG	5	9	-1	-63	3,1
3	7	ME	1	R	1	P	2	4	-1	-28	1,4
2	8	ME	1	R	1	M	3	5	-1	-40	2,0
3	7	ME	1	R	1	P	2	4	-1	-28	1,4
3	9	MA	3	R	1	M	3	7	-1	-63	3,1
1	6	ME	1	R	1	MG	5	7	-1	-42	2,1
3	9	ME	1	R	1	M	3	5	-1	-45	2,2
2	8	ME	1	IR	3	G	4	8	-1	-64	3,2
2	6	ME	1	R	1	MG	5	7	-1	-42	2,1
3	9	MA	3	R	1	M	3	7	-1	-63	3,1
2	8	ME	1	R	1	M	3	5	-1	-40	2,0
3	7	ME	1	R	1	P	2	4	-1	-28	1,4
3	9	MA	3	R	1	M	3	7	-1	-63	3,1
2	7	MA	3	R	1	M	3	7	-1	-49	2,4
3	9	ME	1	R	1	M	3	5	-1	-45	2,2
2	8	ME	1	R	1	M	3	5	-1	-40	2,0
3	9	MA	3	R	1	M	3	7	-1	-63	3,1
2	8	ME	1	IR	3	G	4	8	-1	-64	3,2
3	9	MA	3	R	1	M	3	7	-1	-63	3,1

A.2.32 - Filling the tank	3.03	DI	2	LO	1	IM	2	PE	3	EC
	3.11	DI	2	RE	2	ME	1	PE	3	EC
	3.12	DI	2	LO	1	IM	2	PE	3	EC
	3.19	DI	2	LO	1	IM	2	TE	1	PR
	3.22	DI	2	LO	1	IM	2	PE	3	EC
	3.23	DI	2	LO	1	IM	2	PE	3	EC

A3. Operation phase	Impact	Incidence	Value	Distributivity	Value	Time of Incidence	Value	Length of stay	Value	Probability
A.2.33 - Pre-operational tests	3.10	IN	1	RE	2	ME	1	PE	3	MP
	3.12	DI	2	LO	1	IM	2	PE	3	EC
	3.13	DI	2	LO	1	IM	2	PE	3	EC
A.2.34 - Decommissioning and cleaning up construction sites	3.01	DI	2	LO	1	IM	2	TE	1	PR
	3.02	IN	1	LO	1	ME	1	TE	1	EC
	3.04	DI	2	LO	1	IM	2	TE	1	EC
A3. Operation phase	Impact	Incidence	Value	Distributivity	Value	Time of Incidence	Value	Length of stay	Value	Probability
A.3.01 - Operation of the Rio do Peixe SHPP	3.10	IN	1	RE	2	ME	1	PE	3	MP
	3.11	DI	2	RE	2	ME	1	PE	3	EC
	3.12	DI	2	LO	1	IM	2	PE	3	EC
	3.13	DI	2	LO	1	IM	2	PE	3	EC
	3.19	DI	2	LO	1	IM	2	TE	1	PR
	3.22	DI	2	LO	1	IM	2	PE	3	EC
	3.24	IN	1	LO	1	ME	1	PE	3	EC
A.3.02 - Routine preventive maintenance and upkeep	3.12	DI	2	LO	1	IM	2	PE	3	EC
	3.13	DI	2	LO	1	IM	2	PE	3	EC
	3.22	DI	2	LO	1	IM	2	PE	3	EC
A.3.03 - Corrective Maintenance	3.12	DI	2	LO	1	IM	2	PE	3	EC
	3.13	DI	2	LO	1	IM	2	PE	3	EC
	3.22	DI	2	LO	1	IM	2	PE	3	EC
A4. Deactivation phase	Impact	Incidence	Value	Distributivity	Value	Time of Incidence	Value	Length of stay	Value	Probability
A.4.01 - Reservoir emptying	3.14	DI	2	LO	1	IM	2	TE	1	EC
	3.15	DI	2	LO	1	ME	1	TE	1	PR
	3.16	DI	2	LO	1	IM	2	PE	3	EC

Value	Magnitude	Synergy	Value	Reversibility	Value	Relevance	Value	Importance	Sense	Impact Value	Normalised Impact Value
3	11	ME	1	IR	3	M	3	7	-1	-77	3,8
3	11	MA	3	IR	3	M	3	9	-1	-99	5,0
3	11	MA	3	IR	3	M	3	9	-1	-99	5,0
2	8	ME	1	R	1	MG	5	7	-1	-56	2,8
3	11	ME	1	IR	3	G	4	8	-1	-88	4,4
3	11	ME	1	IR	3	M	3	7	-1	-77	3,8
1	8	ME	1	IR	3	G	4	8	-1	-64	3,2
3	11	MA	3	IR	3	M	3	9	-1	-99	5,0
3	11	MA	3	IR	3	M	3	9	-1	-99	5,0
2	8	ME	1	R	1	M	3	5	-1	-40	2,0
3	7	ME	1	R	1	P	2	4	-1	-28	1,4
3	9	MA	3	R	1	M	3	7	-1	-63	
Value	Magnitude	Synergy	Value	Reversibility	Value	Relevance	Value	Importance	Sense	Impact Value	Normalised Impact Value
1	8	ME	1	IR	3	G	4	8	-1	-64	3,2
3	11	MA	3	IR	3	M	3	9	-1	-99	5,0
3	11	MA	3	IR	3	M	3	9	-1	-99	5,0
3	11	MA	3	IR	3	M	3	9	-1	-99	5,0
2	8	ME	1	R	1	MG	5	7	-1	-56	2,8
3	11	ME	1	IR	3	G	4	8	-1	-88	4,4
3	9	MA	3	IR	3	G	4	10	-1	-90	4,5
3	11	MA	3	IR	3	M	3	9	-1	-99	5,0
3	11	MA	3	IR	3	M	3	9	-1	-99	5,0
3	11	ME	1	IR	3	G	4	8	-1	-88	4,4
3	11	MA	3	IR	3	M	3	9	-1	-99	5,0
3	11	MA	3	IR	3	M	3	9	-1	-99	5,0
3	11	ME	1	IR	3	G	4	8	-1	-88	4,4
Value	Magnitude	Synergy	Value	Reversibility	Value	Relevance	Value	Importance	Sense	Impact Value	Normalised Impact Value
3	9	MA	3	IR	3	M	3	9	-1	-81	4,0

Value	Magnitude	Synergy	Value	Reversibility	Value	Relevance	Value	Importance	Sense	Impact Value	Value of Normalised Impact
2	7	MA	3	IR	3	M	3	9	-1	-63	3,1
3	11	ME	1	IR	3	M	3	7	-1	-77	3,8

Impactful Actions	Impact	Incidence	Value	Distributivity	Value	Time of Incidence	Value	Length of stay	Value	Probability
A.4.02 - Removal of the Dam, Spillway, Escape Channel, Power House and other civil works	3.14	DI	2	LO	1	IM	2	TE	1	EC
	3.15	DI	2	LO	1	ME	1	TE	1	PR
	3.16	DI	2	LO	1	IM	2	PE	3	EC
	3.17	DI	2	LO	1	ME	1	TE	1	PR
	3.18	DI	2	LO	1	IM	2	TE	1	PR

<table>
<tr><td rowspan="2">Impactful Actions</td><td colspan="10">Environmental Component: Physical Environment
C.1.04
Climate and Air Quality</td></tr>
</table>

A2. Construction Phase	Impact	Incidence	Value	Distributivity	Value	Time of Incidence	Value	Length of stay	Value	Probability
A.2.04 - Clearing and suppression of vegetation in areas of direct intervention (permanent structures, construction sites)	4.01	DI	2	LO	1	IM	2	TE	1	EC
A.2.05 - Cleaning and clearing of dump areas	4.01	DI	2	LO	1	IM	2	TE	1	EC
A.2.06 - Setting up construction sites	4.01	DI	2	LO	1	IM	2	TE	1	EC
A.2.07 - Exploitation of borrow areas (clay soils)	4.01	DI	2	LO	1	IM	2	TE	1	EC
A.2.08 - Sand and gravel extraction	4.01	DI	2	LO	1	IM	2	TE	1	EC
A.2.09 - Management of the dumpsites	4.01	DI	2	LO	1	IM	2	TE	1	EC
A.2.10- Management of soil dumps and deposits of other construction materials	4.01	DI	2	LO	1	IM	2	TE	1	EC
A.2.11 - Construction of new accesses and improvements to existing accesses and paths	4.01	DI	2	LO	1	IM	2	TE	1	EC
3	9	MA	3	IR	3	M	3	9	-1	-81	4,0
2	7	MA	3	IR	3	M	3	9	-1	-63	3,1
3	11	ME	1	IR	3	M	3	7	-1	-77	3,8
2	7	ME	1	R	1	MG	5	7	-1	-49	2,4
1	7	MA	3	R	1	M	3	7	-1	-49	2,4

Value	Magnitude	Synergy	Value	Reversibility	Value	Relevance	Value	Importance	Sense	Impact Value	Value of Normalised Impact
3	9	MA	3	R	1	M	3	7	-1	-63	3,1
3	9	MA	3	R	1	M	3	7	-1	-63	3,1
3	9	MA	3	R	1	M	3	7	-1	-63	3,1
3	9	MA	3	R	1	M	3	7	-1	-63	3,1
3	9	MA	3	R	1	M	3	7	-1	-63	3,1
3	9	MA	3	R	1	M	3	7	-1	-63	3,1
3	9	MA	3	R	1	M	3	7	-1	-63	3,1
3	9	MA	3	R	1	M	3	7	-1	-63	3,1
A.2.12 -	4.01	DI	2	LO	1	IM	2	TE	1	EC

Implementation of a sand transhipment structure											
A.2.13 - Construction of a bridge to cross the River do Fish		4.01	DI	2	LO	1	IM	2	TE	1	EC
A.2.14 - Construction of internal site accesses		4.01	DI	2	LO	1	IM	2	TE	1	EC
A.2.15 - Transport of soil, sand and gravel to fronts and construction sites		4.01	DI	2	LO	1	IM	2	TE	1	EC
A.2.16-Operation of the construction site		4.01	DI	2	LO	1	IM	2	TE	1	EC
A.2.17 - Implementation of upstream and downstream cofferdams and river diversion		4.01	DI	2	LO	1	IM	2	TE	1	EC
A.2.18 - Compulsory excavations		4.01	DI	2	LO	1	IM	2	TE	1	EC
A.2.19 - Transport of excavated materials		4.01	DI	2	LO	1	IM	2	TE	1	EC
A.2.21 - Installation of concrete structures		4.01	DI	2	LO	1	IM	2	TE	1	EC
A.2.22 - Construction of the Escape Channel, Powerhouse and Spillway		4.01	DI	2	LO	1	IM	2	TE	1	EC
A.2.23 - Sub-surface drainage of concrete structures		4.01	DI	2	LO	1	IM	2	TE	1	EC
A.2.24 - Anchoring the escape channel		4.01	DI	2	LO	1	IM	2	TE	1	EC
3	9	MA	3	R	1	M	3	7	-1	-63	3,1
3	9	MA	3	R	1	M	3	7	-1	-63	3,1
3	9	MA	3	R	1	M	3	7	-1	-63	3,1
3	9	MA	3	R	1	M	3	7	-1	-63	3,1
3	9	MA	3	R	1	M	3	7	-1	-63	3,1
3	9	MA	3	R	1	M	3	7	-1	-63	3,1
3	9	MA	3	R	1	M	3	7	-1	-63	3,1
3	9	MA	3	R	1	M	3	7	-1	-63	3,1
3	9	MA	3	R	1	M	3	7	-1	-63	3,1
3	9	MA	3	R	1	M	3	7	-1	-63	3,1
3	9	MA	3	R	1	M	3	7	-1	-63	3,1
3	9	MA	3	R	1	M	3	7	-1	-63	3,1
A.2.25 - Draining, pumping and cofferdamming the area between the upstream and downstream cofferdams		4.01	DI	2	LO	1	IM	2	TE	1	EC
A.2.26 - Surface protection of dams		4.01	DI	2	LO	1	IM	2	TE	1	EC

Impactful Actions	Impact	Incidence	Value	Distributivity	Value	Time of Incidence	Value	Length of stay	Value	Probability
K.221 - Sub-surface drainage of jambs	4.01	DI	2	LO	1	IM	2	TE	1	EC
A.2.28 - Electromechanical assembly	4.01	DI	2	LO	1	IM	2	TE	1	EC
A.2.29 - Execution of complementary civil works	4.01	DI	2	LO	1	IM	2	TE	1	EC
A.2.30 - Clearing vegetation and cleaning the flood area	4.01	DI	2	LO	1	IM	2	TE	1	EC
A.2.31- Rehabilitation of local roads and accesses	4.01	DI	2	LO	1	IM	2	TE	1	EC
A.2.34 - Decommissioning and cleaning up construction sites	4.01	DI	2	LO	1	IM	2	TE	1	EC
A3. Operation phase	Impact	Incidence	Value	Distributivity	Value	Time of Incidence	Value	Length of stay	Value	Probability
A.3.01 - Operation of the Rio do Peixe SHPP	4.02	IN	1	RE	2	ME	1	PE	3	EC

Impactful Actions	Environmental Component: Physical Environment C.1.05 Palaeontological Heritage									
A2. Construction Phase	Impact	Incidence	Value	Distributivity	Value	Time of Incidence	Value	Length of stay	Value	Probability
A.2.06 - Setting up construction sites	5.01	DI	2	LO	1	IM	2	PE	3	PR
3	9	MA	3	R	1	M	3	7	-1	-63	3,1
3	9	MA	3	R	1	M	3	7	-1	-63	3,1
3	9	MA	3	R	1	M	3	7	-1	-63	3,1
3	9	MA	3	R	1	M	3	7	-1	-63	3,1
3	9	MA	3	R	1	M	3	7	-1	-63	3,1
3	9	MA	3	R	1	M	3	7	-1	-63	3,1
3	9	MA	3	R	1	M	3	7	-1	-63	3,1
3	9	MA	3	R	1	M	3	7	-1	-63	3,1

Value	Magnitude	Synergy	Value	Reversibility	Value	Relevance	Value	Importance	Sense	Impact Value	Value of Normalised Impact
3	10	MA	3	IR	3	G	4	10	-1	-100	5,0

Value	Magnitude	Synergy	Value	Reversibility	Value	Relevance	Value	Importance	Sense	Impact Value	Value of Normalised Impact
2	10	ME	1	IR	3	MG	5	9	-1	-90	4,5

Impactful Actions	Impact	Incidence	Value	Distributivity	Value	Time of Incidence	Value	Length of stay	Value	Probability
A.2.07 - Exploitation of borrow areas (clay soils)	5.01	DI	2	LO	1	IM	2	PE	3	PR
A.2.08 - Sand and gravel extraction	5.01	DI	2	LO	1	IM	2	PE	3	PR

A.2.11 - Construction of new accesses and improvements to existing accesses and paths	5.01	DI	2	LO	1	IM	2	PE	3	PR
A.2.12 - Implementation of a sand transhipment structure	5.01	DI	2	LO	1	IM	2	PE	3	PR
A.2.13 - Construction of a bridge to cross the River Peixe	5.01	DI	2	LO	1	IM	2	PE	3	PR
A.2.14 - Construction of internal site accesses	5.01	DI	2	LO	1	IM	2	PE	3	PR
A.2.17 - Implementation of upstream and downstream cofferdams and river diversion	5.01	DI	2	LO	1	IM	2	PE	3	PR
A.2.18 - Compulsory excavations	5.01	DI	2	LO	1	IM	2	PE	3	PR
A.2.32 - Filling the tank	5.01	DI	2	LO	1	IM	2	PE	3	PR

DI = Direct Impact; LO = Local Impact; IM = Immediate Impact; MA = Major Synergy; ME = Minor Synergy; ME = Medium Impact; PE = Permanent Impact; CE = Certain Impact; PR = Probable Impact;
IR = Irreversible Impact; R= Reversible Impact; G = Great Relevance; NE = Negative Sense

Prepared by: Edson Alves Filho.

2	10	ME	1	IR	3	MG	5	9	-1	-90	4,5
2	10	ME	1	IR	3	MG	5	9	-1	-90	4,5
2	10	ME	1	IR	3	MG	5	9	-1	-90	4,5
2	10	ME	1	IR	3	MG	5	9	-1	-90	4,5
2	10	ME	1	IR	3	MG	5	9	-1	-90	4,5
2	10	ME	1	IR	3	MG	5	9	-1	-90	4,5
2	10	ME	1	IR	3	MG	5	9	-1	-90	4,5
2	10	ME	1	IR	3	MG	5	9	-1	-90	4,5
2	10	ME	1	IR	3	MG	5	9	-1	-90	4,5

As for the Relief and Soils Subcomponent, the impacts with the highest values are, respectively:

- 3.11 - Change in humidity conditions due to increase in local water mass (-99),

- 3.12- Increased erosion capacity on the banks of the reservoir due to waves crashing into it and the water table rising (-99),

- 3.13- Change in the degree of criticality and occurrence of gravitational mass processes on the banks of the reservoir (-99) e;

- 3.24 - Formation and Perennialisation of Wetlands and Flooded Areas (-90).

Evidence of these impacts was analysed both through the assessment of geo-indicators from the consolidated intervention phase (**Section 2.4** of this volume) and through the historical mapping carried out, which took into account the pre-intervention, active intervention and consolidated intervention phases (**Sections 2.2, 2.3 and 2.4** of this volume), where it was possible to delimit the emergence of new flooded areas and processes related to slope slides in the reservoir area.

When analysing the Climate and Air Quality Subcomponent, only one impact with a significant value is mentioned, relating to impact 4.02 - Local Climate Change (-100). Once again, due to the non-availability of meteorological data on a local scale in the documents consulted, it was not possible to verify possible changes in the local climate within the geoindi- cators selected. The available data only refers to projections of climatic parameters taken by measuring stations located in basins close to the development.

Finally, the last sub-component analysed in the integrated environmental assessment of the physical environment for the study area refers to palaeontological heritage. It should be noted that the inclusion of palaeontological studies in environmental impact assessments is recent, which justifies the lack of consideration of this variable in the studies consulted. Only one impact was considered for this sub-component, called 5.01 - Risk of Loss of Palaeontological Sites (-90). It was not possible to establish any kind of verification of the effectiveness of this impact in the study area due to the lack of data of this nature, which makes the inclusion of this impact a potential one for the study area. **Map 2.1.m** - Negative Environmental Impacts on the Physical Environment in the study area is presented below, derived from the spatialisation and quantification of the impacts presented in **Table 2.1.a**.

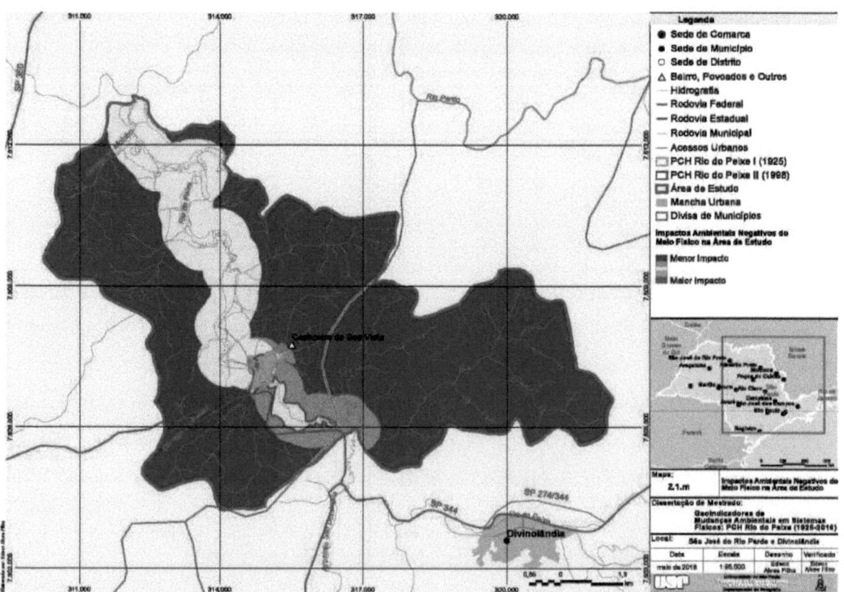

Map 2.1.m - Negative Environmental Impacts of the Physical Environment in the study area

Map 2.1.m shows that most of the negative impacts listed so far for the study area are confined to the reservoir area and its surroundings, where suppression activities to clear the flood basin and open access roads take place during the implementation phase. Exceptions are the areas located along the dividing lines in the southwest and north, where the steep slopes associated with the granitoid and charnokite areas are favourable to the development of erosion processes. With regard to the silting up of watercourses, the greatest impact is mentioned in the area around the reservoir and the downstream stretch of the River Peixe from the dam to the confluence with the River Pardo.

The Integrated Environmental Assessment of the Physical Environment for the study area concludes with the presentation of **Map 2.1.n** - Environmental Fragility of the Physical Environment in the study area, created from the spatial intersection between Maps 2.1.j (Synthetic Sensitivity of the Physical Environment in the Study Area) and 2.1.l (Environmental Impacts of the Physical Environment in the Study Area), assigning equal weights to the two categories (0.5). After spatially intersecting the *sdapefiie* files of the two maps, a column was created in the database with the weight value, and the normalised sensitivity and negative impact values were multiplied with these weights to obtain the fragility values for the study area. A choropleth scale was then applied to the fragility values.

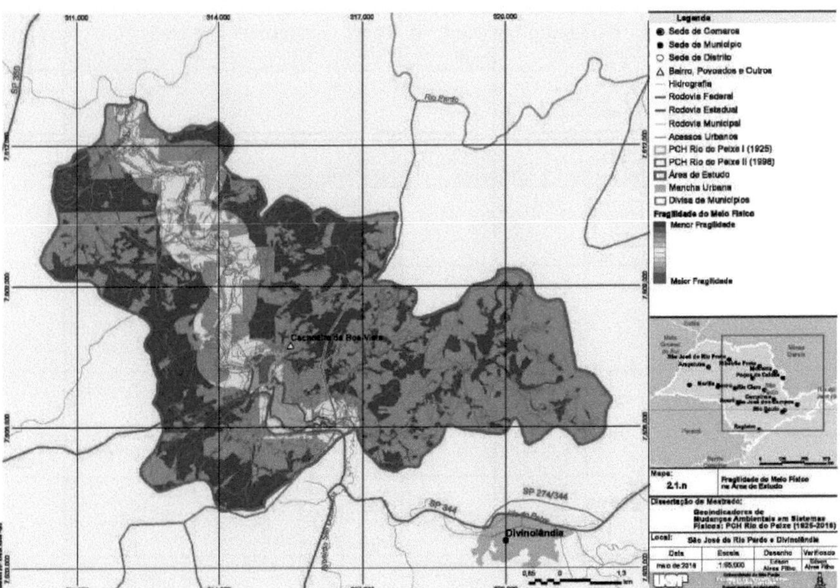

Map 2.1.n - Environmental Fragility of the Physical Environment in the study area

The Physical Environment Fragility Map for the Study Area was composed from the intersection of the Physical Environment Sensitivity Map and the Physical Environment Negative Impacts Map.

As already mentioned, in the case of Sensitivity, a strip stretching from the north-west to the south-east of the study area presents the greatest limitations to the physical environment.

Looking at Map 2.1.n, the areas classified as most fragile are confined to the immediate surroundings of the reservoir and civil structures, affected by the Rio do Peixe I and II SHPs. With a slightly lower degree of fragility, the southwestern portion is mentioned, where the limitations of soil erosion and geology are less favourable. With a medium degree of fragility, the southeast portion is mentioned, where there are areas of elongated hills, steep slopes and sloping areas.

2.2 Application of Geoindicators relating to Semi-Preserved Original Morphology (Pre-Intervention Phase)

The application of the geo-indicators and the quantitative analyses derived from them were based on the Topographic Chart of the Geological and Geographical Institute (1905), 1930 aerial photographs from the São Paulo State Energy Foundation and the mapping of the Primitive Vegetation Cover of the State of São Paulo (1920) drawn up by Vitor. *et. al.* (1975).

Among the geo-indicators systematised in Table 2.2.a, those of a morphological/ morphometric nature were applied to the condition of semi-preserved original morphology, as well as information derived from land use and occupation (areas of earth movement/exposed soil, land use categories).The results obtained for the semi-preserved original morphology geo-indicators applied to the study area are shown in **Table 2.2.a**:

Table 2.2.a - Geoindicators of Semi-Preserved Original Morphology applied to the Pre-Intervention Phase (1930)

System	Subsystem	Indicator	Parameters	Results
Basin Hydrographic	-	Shape	Drainage pattern of the catchment area	Dendritic to Subdendritic
Basin Hydrographic	-	Shape	Length of the River Basin's surface course (Eps)	150 m
Basin Hydrographic	-	Dividers	Number of Dividers	50
Basin Hydrographic	-	Dividers	Divider Line Extension	52.41 kilometres
Strand/Fluvial/Lake System	Plains/Terraces/Canal/Reservoir	Shape	Length Channel Segments (C.s.)	171.56 kilometres
Basin Hydrographic		Area	Drainage Area	53.89 km^2
Basin Hydrographic		Density	Hydrographic density of the catchment area (Dh)	4.60 rivers/km^2
Basin		Density	Drainage	3.13 km/km^2

Hydrographic			Density of the Basin Hydrographic (Dd)	
Basin Hydrographic		Density	River Basin Maintenance Coefficient (Cm)	310 m /m^2
Basin Hydrographic		Vegetation cover	Type	Montane Semideciduous Seasonal Forest (27.70 km^2), Semideciduous Seasonal Forest Alluvial (1.85 km)^2
Basin Hydrographic		Vegetation cover	Internships Succession	Analyses of 1930 aerial photographs showed that the study area was covered by Montane Semideciduous Seasonal Forest in an Advanced Stage of Regeneration, as well as Alluvial Semideciduous Seasonal Forest in an Advanced Stage of Regeneration. Regeneration
Basin Hydrographic		Vegetation cover	Volume	Montane Semideciduous Seasonal Forest (149.58 m^3 /ha), Semideciduous Seasonal Forest Alluvial (10.0 m^3 /ha)
Basin Hydrographic		Daily Precipitation	mm/day	8.91 mm/day (January - wettest month)
Basin Hydrographic		Annual rainfall	mm/year	1,201.9 mm/year
Basin Hydrographic		Area of Occupation of Slope Classes	km /class^2	4.97 km^2 (up to 2%), 5.23 km^2 (2% to 8%), 10.82 km^2 (8% to 15%), 23.88 km^2 (15% to 30%), 8.66 km^2 (30% to 45%), 1.14 km^2 (> 45%)
Strand/ River/Lak e System	Plains/Terraces/Cana l/Reservoir	Number of ruptures	unit	68
Strand/ River/Lak e System	Plains/Terraces/Cana l/Reservoir	Area of Occupation of Slope Classes	km /class^2	2.67 km^2 (up to 2%), 0.74 km^2 (2% to 8%), 0.57 km^2 (8% to 15%)
Strand/ River/Lak e System	Plains/Terraces/Cana l/Reservoir	Average altitude of the slopes	m	950 m
Strand/ River/Lak e System	Plains/Terraces/Cana l/Reservoir	Guidance Prevailing in the Vertentes	direction	4.96 km^2 (flat), 6.16 km^2 (north), 4.22 km^2 (north-east), 5.50 km^2 (east), 6.02

				km^2 (south-east), 5.39 km^2 (south), 4.46 km^2 (south-west), 8.52 km^2 (west), 9.47 km^2 (north-west),
Strand/ River/Lak e System	Plains/Terraces/Cana l/Reservoir	Average Ramp Length	m	2.100 m
Strand/ River/Lak e System	Plains/Terraces/Cana l/Reservoir	Extension of the Rupture Lines	km	45.79 kilometres
Strand/ River/Lak e System	Plains/Terraces/Cana l/Reservoir	Number of changes	unit	288
Strand/Fluvial /Lake System	Plains/Terraces/Cana l/Reservoir	Extension of Moving Lines	km	156.07 km
Strand/ River/Lak e System	Plains/Terraces/Cana l/Reservoir	Morphologies Predominant in Plan and Profile	description	5.41 km^2 (tops), 2.62 km^2 (river plains), 0.93 km^2 (river terraces), 2.46 km^2 (broad concave amphitheatres), 43.28 km^2 (slopes with an indiscriminate concave and convex profile).
Strand/ River/Lak e System	Plains/Terraces/Cana l/Reservoir	Average length of slopes	km	2.1 kilometres
Strand/ River/Lak e System	Plains/Terraces/Cana l/Reservoir	Degree of Vertical Dissection of Occupied Areas		Average of 30 metres
Strand/Flu vial/Lake System	Plains/Terraces/Cana l/Reservoir	Degree of Horizontal Dissection of Occupied Areas		Average 300 metres
Basin Hydrographic	-	Aquifer type (fractured, porous)		Crystalline (fractured)
Basin Hydrographic	-	Units Geological	km^2	10.95 km^2 (migmatites of the Varginha- Guaxupé Group), 16.12 km^2 (paragneisses of the Varginha-Guaxupé), 27.61 km^2 (charnockites of the São José do Rio Pardo-Divinolândia Group).
River System	Plains/Terraces/Cana l/Reservoir	Total Area of Water Bodies	ha	20 ha
River System	Plains/Terraces/Cana l/Reservoir	Total Perimeter of Water Bodies	m	14.127 m
System River	Plains/Terraces/Cana l/Reservoir	River Plain Area	ha	262 ha
River System	Plains/Terraces/Cana l/Reservoir	Perimeter River Plain	km	70.48 kilometres
System River	Plains/Terraces/Cana l/Reservoir	Distribution of the River Plain	description	It is distributed in the study area in a NO-SE direction

System River	Plains/Terraces/Canal/Reservoir	River Terrace Area	ha	93 ha
River System	Plains/Terraces/Canal/Reservoir	Perimeter of River Terraces	m	16.711 m
River System	Plains/Terraces/Canal/Reservoir	Distribution of River Terraces	description	It is distributed in two bands in the study area: one to the NE, downstream of the Rio do Peixe I and II SHPs, and one to the S, just upstream of the Rio do Peixe I and II SHPs.
System River	Plains/Terraces/Canal/Reservoir	Distribution of River Valleys	description	They are widespread throughout the study area, but with higher densities in the north-west and north-east.
River System	Plains/Terraces/Canal/Reservoir	Extension of the River Valleys	m	104.41 kilometres
River System	Channel	Main Channel Flow (Q7, Q10)	m^3/s	$8.34\ m^3/s$
System River	Channel	Peak Flow	m^3/s	$73.50\ m^3/s$
River System	Channel	Original Channel Extension	km	19.32 km (Peixe River)
System River	Channel	Channel Section Width (Wmp)	m	22,50 m
River System	Channel	Surface width (L)	m	12,25 m
River System	Channel	Average Section Depth (Dmp)	m	2,20 m
System River	Channel	Longitudinal Channel Profile	description	Characteristic equilibrium profile with knick *points* at the confluence with higher order channels.
River System	Channel	River Gradient	m	24,52 m
River System	Channel	Cross-sectional area (Amp)	m^2	$19,31\ m^2$
System River	Channel	Wet Area (A)	m^2	$19,28\ m^2$
River System	Channel	Wet Perimeter (P)	m	26,90 m
System River	Channel	Hydraulic Radius (R)	m/m^2	$0.71\ m/m^2$
System River	Channel	Channel Shape Index (F)		0,26
System River	Channel	Original bed sinuosity index		2,50

Prepared by: Edson Alves Filho

With regard to the morphometric parameters of the river basin system, taken from the mapping of the semi-preserved original morphology of the study area, the dendritic drainage pattern is observed, with the main course of the Peixe river orientated in a NO-SE direction, with the tributaries on the right bank, such as the Boa Vista stream, orientated NE-SO. On the left bank, the main tributaries, such as the Cachoeira and Moinho streams, are orientated N-S and SO-NE.

The length of the basin's surface course is 150 m, followed by a total length of river channels (perennial and intermittent) of around 171.56 km. With regard to hydrographic density, the values found were 4.60 rivers/km^2 , while for drainage density a total of 313 km/km was obtained.

The drainage area represented by the study area is 53.89 km^2 , with a maintenance coefficient of 310 m^2 /m. With regard to the hydrographic dividers, 50 dividers were counted in the condition of semi-preserved original morphology, with a length of 52.41 km.

With regard to vegetation cover and other land use categories, there is a predominance of areas related to Montane Seasonal Semideciduous Forest in an advanced stage of regeneration, representing 45% (23.35 km^2) of the study area. Next,
appear areas of Alluvial Semideciduous Seasonal Forest in an advanced stage of regeneration, representing 16% (8.6 km^2) of the study area. This is followed by areas of permanent crops, with 25% (13.47 km^2) of the study area, and finally temporary crops, with 14% (7.54 km^2) of the study area. The distribution pattern of the vegetation cover shows that the masses of Montane Semideciduous Seasonal Forest are more representative in the sectors with slopes of more than 30%, while the areas of Alluvial Semideciduous Seasonal Forest mainly accompany the river plain and the terraces of the River Peixe. The areas of permanent crops develop along farms and ranches, as well as in the middle third of the slopes, and finally, the areas with temporary crops are located in the medium and low slope sectors, where the slope intervals are less than 15%.

In order to estimate the volume of vegetation cover and other land use categories found in the study area, estimates were made of the average diameter and height of the different vegetation types in forest inventory manuals (FUNDAÇÃO FLORESTAL, 2008). This resulted in a forest volume of 149.58 m^3 /ha for the areas with Montane Semideciduous Seasonal Forest, and 10.0 m^3 /ha for the areas with Alluvial Semideciduous Seasonal Forest.

For the climatic parameters relating to the basin-hydrographic system, the Environmental Impact Study for the Rio do Peixe II SHPP (CEMA, 1993) found historical series for the study area with data prior to the implementation of the Rio do Peixe I SHPP, showing an average daily rainfall of 8.91 mm/day, based on the month of January, the wettest in the entire series considered (1925-1998). In terms of annual rainfall, 1,201.9 mm/year was found for the same period.

With regard to the distribution of slope classes throughout the river basin system, the areas located in slope intervals of less than 2% correspond to 4.97 km^2 . The areas between 2% and 8%

correspond to 5.23 km^2 , followed by an area of 10.82 km^2 for the slope interval between 8% and 15%. Finally, the areas with the steepest slopes are distributed as follows: 23.88 km^2 , for the 15% to 30% interval, 8.66 km^2 for the areas between 30% and 45% and 1.14 km^2 for the areas with slopes greater than 45%.

When analysing the lithostratigraphic units that support the landscapes in the study area, there is a predominance of chamoquites from the São José do Sul Group.

Rio Pardo - Divinolândia, with an area of 27.61 km^2 , which represents 53.20% of the total area. Next are the Paragnaisses of the Varginha - Guaxupé Group, with an area of 16.12 km^2 and representing 29.21% of the total analysed. Finally, the migmatites of the Varginha - Guaxupé Group totalled 10.16 km^2 or 17.59% of the study area. As for the aquifer units, the entire study area (53.89 km^2) corresponds to the crystalline aquifer.

When the geo-indicators relating to the slope system are analysed in relation to the morphographic elements with semi-preserved original morphology, the situation is as follows: - Main col: 47 cols were mapped, with a length of 11.46 kilometres.

- Secondary colonnade: 9 colonnades were mapped, spanning 5.94 kilometres.
- Main Divider: 9.92 kilometres of main dividers were mapped.
- Secondary Divider: 8.48 kilometres of secondary dividers were mapped.
- Concave Slope Ruptures in the upper third of the slope: totalling 4.80 km in length.
- Concave Slope Ruptures in the Middle Third of the Slope: totalling 5.26 km in length.
- Concave Slope Ruptures in the lower third of the slope: totalling 2.13 km in length.
- Convex slope ruptures in the upper third of the slope: totalling 4.1 km in length.
- Convex Slope Ruptures in the Middle Third of the Slope: totalling 1.56 km in length.
- Convex slope ruptures in the lower third of the slope: totalling 2.4 km in length.
- Concave changes of slope in the upper third of the slope: totalling 2.36 km in length.
- Concave changes of slope in the middle third of the slope: a total of 3.25 kilometres.
- Concave changes of slope in the lower third of the slope: totalling 1.18 km in length.
- Convex changes of slope in the upper third of the slope: totalling 1.12 km in length.
- Convex changes of slope in the middle third of the slope: totalling 1.8 km in length.
- Convex slope changes in the lower third of the slope: totalling 2.16 km in length.
- Defined Slope-Plain Rupture: totalling 0.9 km in length.
- Rupture Slope-Undefined Plain: totalling 0.6 km in length.
- V-shaped valleys: totalling 120.6 km in length.
- Cradle Valleys: 50.9 kilometres long.

The average length of the slopes is 2.1 kilometres, while the altitude is 950 metres. When observing the degree of vertical dissection of the occupied areas, values of 30 m are found, while for horizontal dissection, the average values found are 300 m.

The predominant orientation found on the slopes is 9.47 km^2 for the north-west face, 8.52 km^2 for the west face, 6.02 km^2 for the south-east face, 5.39 km^2 for the south face, 4.96 km^2 located in areas with no orientation, and therefore flat terrain, 4.46 km^2 for slopes with a south-west face, and finally 4.22 km^2 for slopes with a north-west orientation. The average ramp length found was 2,100 metres.

When analysing the geo-indicators relating to the canal subsystem, especially the main canal flows and peak flows, the data taken from the EIA-RIMA for the Rio do Peixe II SHPP gives values of 8.34 m3/s for the first variable and 73.50 m^3 /s for the second variable.

The total area of the water bodies found along the plains/terraces subsystem was 20 ha, while their perimeter was 14.12 km.

For the river plain of the Peixe river, in the condition of semi-prepared original morphology, an area of 262 ha was found, while its perimeter was 70.48 metres. The distribution of this river plain follows a NO-SE orientation.

The river terraces covered an area of 93 ha and had a perimeter of 16.71 km. The distribution of the terraces followed a NE and S orientation.

Still in relation to the geoindicators listed for the channel subsystem, the length of the original channel, the width of the main channel section and the average depth of the section were 19.32 km, 22.50 m and 2.20 m respectively. The surface width of the main channel was 12.25 metres. The longitudinal profile of the channel was classified as being in an equilibrium condition, with the presence of *knickpoints* at the confluence of the higher order channels. The fluvial gradient found was 24.52 metres. The calculated cross-sectional area and wetted area of the main channel were 19.31 m^2 and 19.28 m^2 , respectively. The calculated wetted perimeter was 26.90 metres.

With regard to river geometry, a hydraulic radius of 0.71 m/m was found[2] and a shape index and sinuosity of the original bed of 0.26 and 2.50 respectively.

The geo-indicators relating to Anthropogenic Morphology in the Active Intervention Phase are presented below.

In addition, in order to spatialise the geo-indicators relating to the Semi-Preserved Original Morphology in the Pre-Intervention stage, allowing their subsequent comparison with the Methodology for the Integrated Environmental Assessment of the Physical Environment of Hydroelectric Projects (Item 8.5), **Map 2.2.a** - Semi-Preserved Original Morphology in the Pre-Intervention Stage in the Study Area was drawn up.

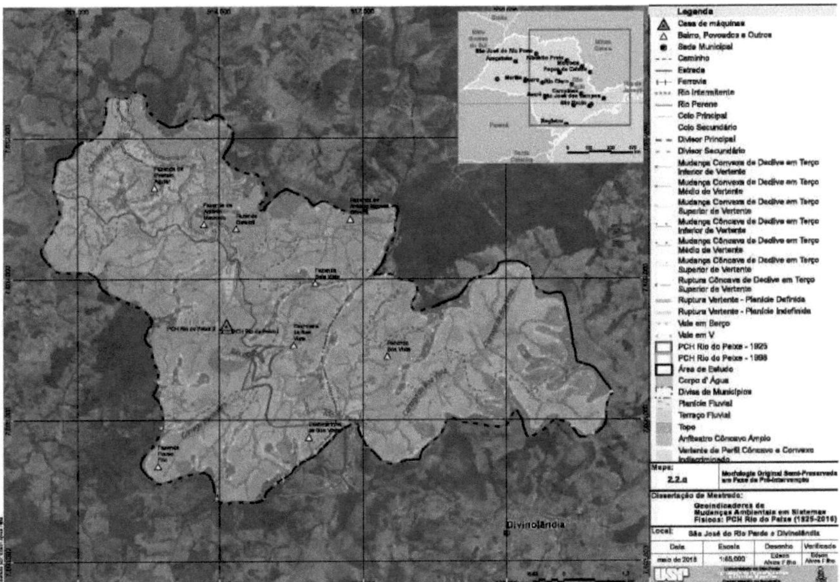

Map 2.2.a - Semi-Preserved Original Morphology in the Pre-Intervention Phase in the Study Area

In relation to the linear elements of Map 2.2.a, referring to the morphological and morphometric elements of the slopes and dividers, it can be seen that the main dividers are located at the top, immediately at the headwaters of the main drainages in the study area, such as the Moinho and Cachoeira streams, on the left bank of the Peixe river, and the Boa Vista stream, on the right bank of the Peixe river.

The boundary of the main dividers is delimited by the main hills, which are also located at the top immediately at the headwaters of the aforementioned drainages. The secondary dividers are located in the north, centre and south-west of the study area, near the headwaters of the Cachoeira stream and along the headwaters of some tributaries of the Peixe river on the right bank. The boundary of the secondary dividers is given by the secondary colons, located in the same positions as those described for the secondary dividers.

The concave slope changes of the upper third of the slope are located in the north-west, south-west and east of the study area, occupying the headwaters along the Moinho, Cachoeira and Boa Vista streams. Next, the concave slope changes of the middle third of the slope are concentrated in the north-west and south-west of the study area, where there are areas of small hills, hillsides and hills with markedly concave profiles limited by subtle changes in slope, where agricultural use was sought in the study area. Finally, the concave slope changes in the lower third of the slope are markedly

45

concentrated in the north-west and south-west portions of the study area, along the low slope sectors drained by the Moinho and Cachoeira streams and their tributaries, close to their respective confluences with the Peixe river.

When analysing the convex changes in slope, located in the upper third of the slopes, there is a generalised distribution throughout the study area, concentrated near the main and secondary dividers and colons. The convex changes in slope located in the middle third of the slopes are also spread throughout the study area, with the exception of the eastern portion of the study area, near the headwaters of the Boa Vista stream. Finally, the convex changes of slope in the lower third of the slopes are preferentially positioned in the north-west and south-west portions of the study area, near the confluence of the Moinho, Cachoeira and tributary streams with the Peixe river.

Concave breaks were only found in the study area in the upper third of the slopes, mainly in the central and south-western portions of the study area, close to the headwaters of the Cachoeira stream. Convex slope breaks were also only found in the study area in the upper third of the slopes, located near the dividers in the north-west and north-east of the study area.

The defined slope-plain breaks follow the course of the Peixe River in the study area, travelling in an S - NO direction, where the terrace levels are more pronounced and easily discernible in relation to the river plain. Finally, the undefined slope-plain breaks follow the same S - NO direction mentioned, but in sectors where the junction with the slopes is so gentle, due to the little development of the river terraces, that it is not possible to estimate a boundary during the stereoscopic restitution procedure.

Finally, the cradle valleys spread throughout the study area, but occupy segments of their courses located in the middle and lower thirds of the slopes, while the V valleys are concentrated in the N, NO, NE, SO and S portions of the study area, with their courses located in the upper thirds of the slopes.

With regard to the areal elements of the Map of the Original Semi-Preserved Morphology of the Study Area, mention should be made of the areas of tops, scattered throughout the study areas, accompanying the areas of main and secondary dividers and hillocks. Areas of broad concave amphitheatres are concentrated in the eastern part of the study area, in the region of the headwaters of the Boa Vista stream. A second stretch of broad concave amphitheatres is located in the south-west of the study area, in the region of the headwaters of the Cachoeira stream. The area of indiscriminate concave and convex slopes is generally located throughout the study area, bordering the tops - sectors of the plain and terraces of the Peixe river. Finally, the areas of plains and river terraces follow the course of the River Peixe in a S - NW direction.

2.3 Application of Geoindicators related to Anthropogenic Morphology

The application of the geo-indicators for the active intervention phase was based on the mapping of physical land use categories, as well as the geo-indicators discernible in the IKONOS satellite image dated 1998, as well as the Environmental Impact Study (EIA-RIMA) for the Rio do Peixe II SHPP and the Report on the Renewal of the Operating Licence for the Rio do Peixe II SHPP.

Firstly, the quantifications and descriptions of the evolution of land use are presented, with the time frame from 1930, taking as a starting point a date that includes the condition of semi-preserved original morphology. Secondly, the evolution of land use is presented for the period between 1977 and 1983, which covers a period of interventions in the landscape that already includes the implementation of the Rio do Peixe I SHPP and precedes the implementation of the Rio do Peixe II SHPP, still within the Semi-Preserved Original Morphology Phase. Thirdly, the land use for the year 1998 is presented, which is used to describe the pattern of morphological interventions for the active intervention phase, and finally, in a fourth moment, the maps for the years 2006 and 2016 are presented, which mark the period between the active and consolidated intervention phases.

To map land use in the condition of semi-preserved original morphology, as already mentioned, we used the 1930 Aerial Photographs from the São Paulo State Energy Foundation, as well as the Casa Branca Sheet, a topographic map on a scale of 1:100,000, drawn up by the Geological and Geographical Institute (IGC, 1905) on which we observed the contour lines and toponyms, important indications of land use in the period. In order to complement the boundaries between native vegetation cover and anthropogenic uses, we consulted the mapping of primitive vegetation in the state of São Paulo, drawn up by Vitor eí. *a(.* (2005), on a scale of 1:200,000. The mapping was also complemented by consulting geography theses for this region, where it was possible to find chapters dedicated to land use and the process of territorial occupation of the entire study region.

From this mapping (**Map 2.3.a**) it can be seen that a large part of the study area is covered by masses of Montane Semideciduous Seasonal Forest, with a percentage of 42.06%. Next in line is the Alluvial Semideciduous Seasonal Forest, which occupies around 6.20% of the study area. Herbaceous cover comes next, occupying around 18.42% of the study area, which demonstrates the early development of dairy and beef cattle in the area in question. Agricultural areas (permanent and temporary crops) account for a total of 5.07% of the study area and rural centres for around 0.22%.

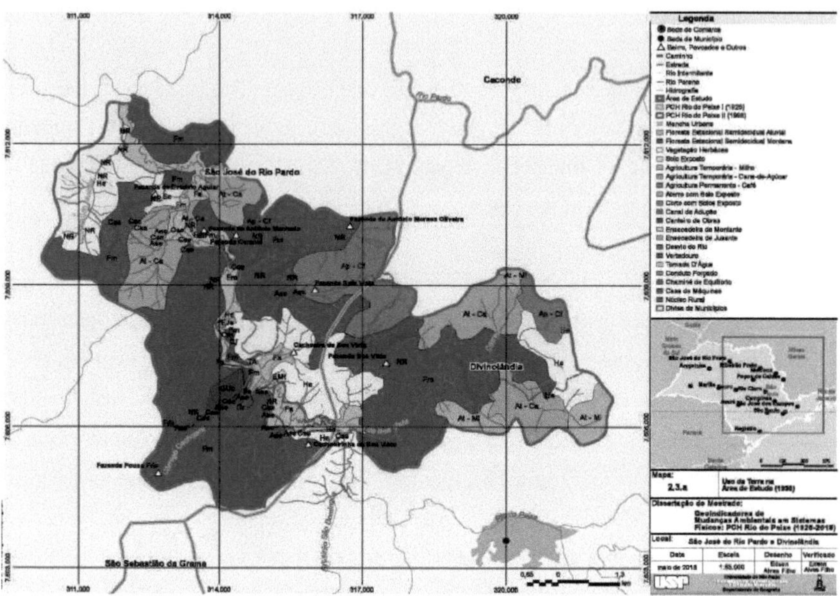

Map 2.3.a - Land Use and Occupation in the Study Area (1930)

Many of the agricultural areas and rural centres indicated in this mapping have been preserved as such to this day, which corroborates the historical descriptions of the process of occupation of the study area, strongly marked by immigration and dedication to farming.

The mapping of land use and associated morphological interferences for 1972 (**Map 2.3.b**) was carried out, as already mentioned, from the restitution of aerial photographs 18038,18039,18040,18041,18042,18043 (Track 20), 26518,26519, 26520, 26521, 26522, 26523, 26541, 26542, 26543, 26544, 26545 and 26546 (Track 58), on a scale of 1:25,000, supplied by Base Aerolevantamentos S.A..

Based on this mapping, it is possible to see a predominance of herbaceous vegetation in the study area, with a percentage of occupied area of 23.08%. This is followed by areas of Montane Seasonal Semideciduous Forest, with 10.11% of the area in question. Agricultural crops occupy third place, with 7.34% of the total. Alluvial Semideciduous Seasonal Forest corridors follow, with 2.01% of the study area. Areas with hygrophilous vegetation, bamboo groves and groves appear next, with 0.59%, 0.004% and 0.20% of the study area, respectively. Finally, areas of cuttings, landfills and rural centres complement the land use categories for the year in question, with percentages of 0.13%, 0.07% and 0.26% respectively.

48

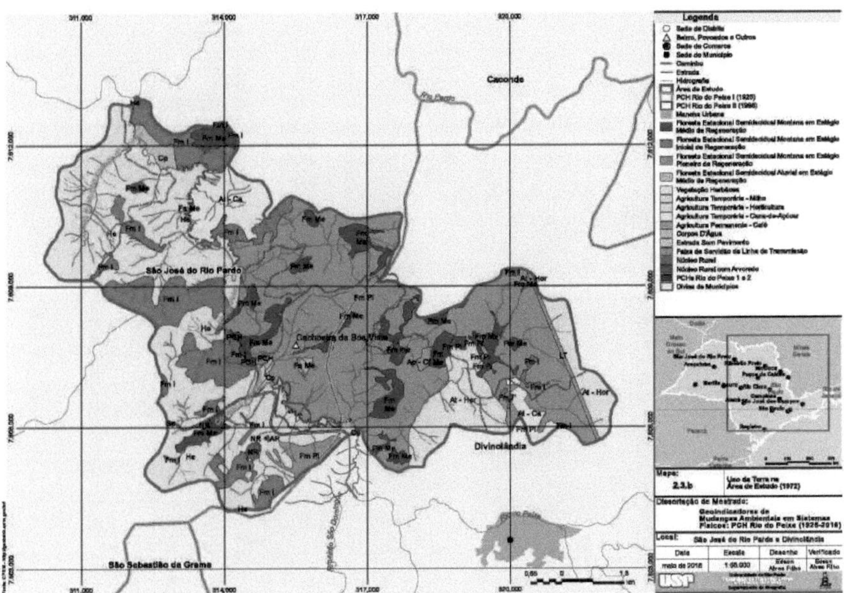

Map 2.3.b - Land Use and Occupation in the Study Area (1972)

For the mapping of land use and morphological interference in 1983 (Map 2.3.c), as already mentioned, aerial photographs 10, 11, 12 (Track 33 B), 41, 42, 43, 44, 45, 46 (Track 34) were used, on a scale of 1:35,000, acquired from the company Base Aerolevantamentos S.A.

Based on this mapping, it is possible to see a predominance of herbaceous vegetation in the study area, with an occupied area percentage of 33.83%. This is followed by areas of Montane Seasonal Semideciduous Forest, with 33.44% of the area in question. Agricultural crops occupy third place, with 8.28 per cent of the total. Next up are corridors of Alluvial Semideciduous Seasonal Forest, with 2.07% of the study area. Areas with hygrophilous vegetation, bamboo groves and groves appear next, with 0.94%, 0.007% and 0.26% of the study area, respectively. Finally, areas of cuttings, landfills and rural centres complement the land use categories for the year in question, with percentages of 0.20%, 0.09% and 0.28% respectively.

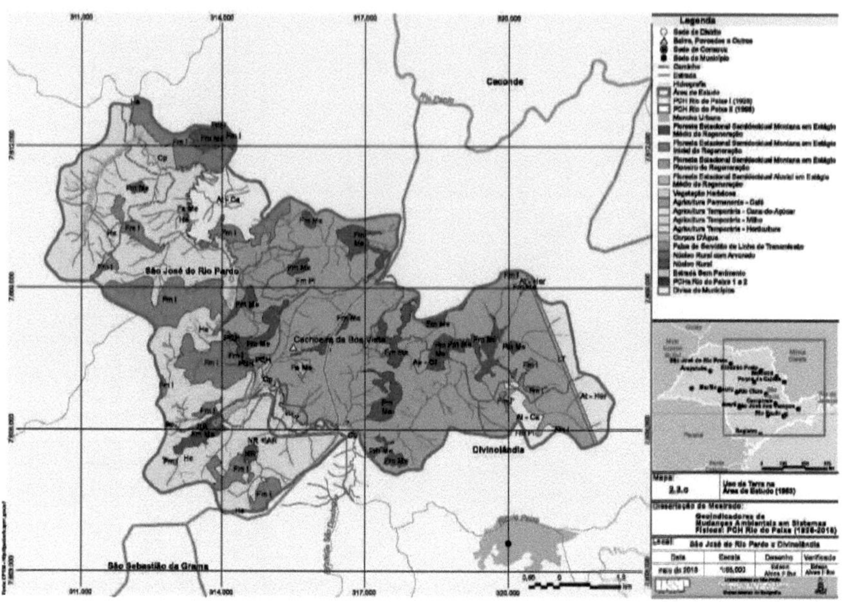

Map 2.3.c - Land Use and Occupation in the Study Area (1983)

The 1998 land use mapping (**Map 2.3.d**) used, as already mentioned, the Map of Vegetation Cover and Land Use in the Area of Direct Influence of the Rio do Peixe II SHPP, drawn up by CEMA Consultoria for the Environmental Impact Study (EIA-RIMA) for this project.

Based on this mapping, it is possible to see a predominance of herbaceous vegetation in the study area, with an occupied area percentage of 42.92%. This is followed by areas of Montane Seasonal Semideciduous Forest, with 40.35% of the area in question. Agricultural crops occupy third place, with 9.34% of the total. Alluvial Semideciduous Seasonal Forest corridors follow, with 2.71% of the study area. Areas with hygrophilous vegetation, bamboo groves and groves appear next, with 1.22%, 0.007% and 0.30% of the study area, respectively. Finally, areas of cuttings, landfills and rural centres complement the land use categories for the year in question, with percentages of 0.37%, 0.15% and 0.29% respectively.

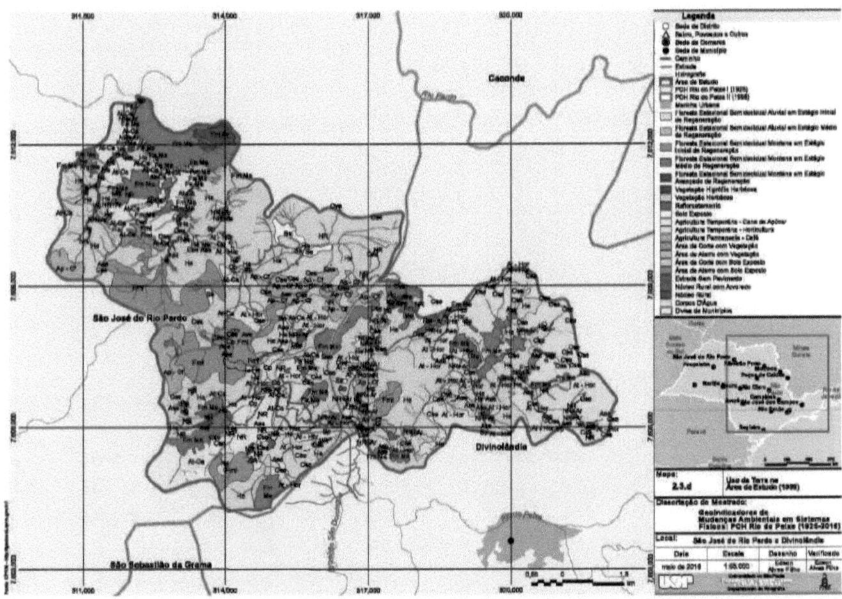

Map 2.3.d - Land Use and Occupation in the Study Area (1998)

For the 2006 land use mapping (**Map 2.3.e**), as already mentioned, aerial photographs 8524, 8525, 8526, 8527 (Track 152), 8640, 8641, 8642, 8643, 8644 (Track 153), 8665, 8666, 8667 (Track 154), acquired from the company Base Aerolevantamentos S.A., were used.

Based on this mapping, it is possible to see a predominance of herbaceous vegetation in the study area, with an occupied area percentage of 41.72%. This is followed by areas of Montane Seasonal Semideciduous Forest, with 33.62% of the area in question. Agricultural crops occupy third place, with 9.91 per cent of the total. Next up are corridors of Alluvial Semideciduous Seasonal Forest, with 2.93% of the study area. Areas with hygrophilous vegetation, bamboo groves and groves appear next, with 0.85%, 0.01% and 0.33% of the study area, respectively. Finally, areas of cuttings, landfills and rural centres complement the land use categories for the year in question, with percentages of 0.38%, 0.18% and 0.36% respectively.

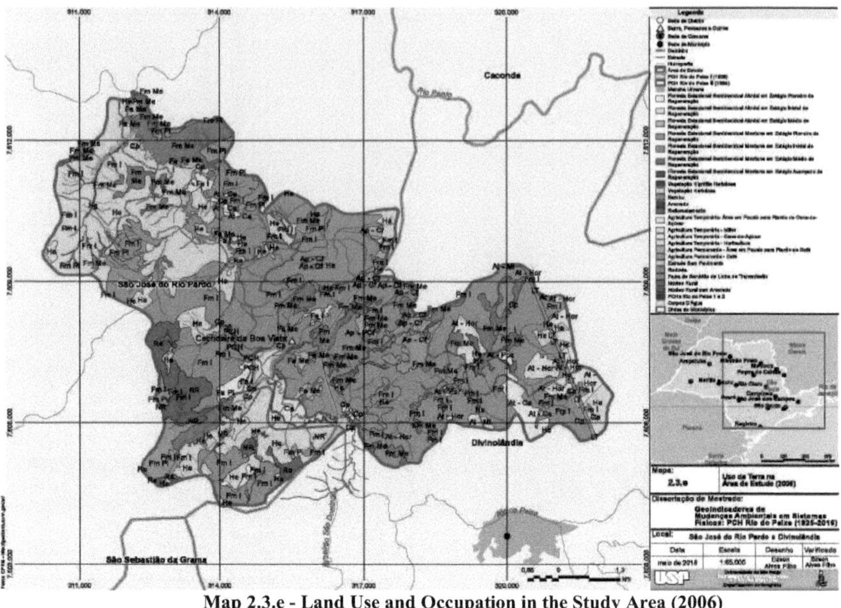

Map 2.3.e - Land Use and Occupation in the Study Area (2006)

Finally, for the 2016 land use mapping (Map 6.7.b, Volume I), Google Earth images dated June were used, with a spatial resolution of less than 1 metre, which is therefore compatible with mapping on a scale of 1:10,000 or greater. Using the *SAS Planet* software, an area covering one and a half times the study area was captured using a polygon, with a scan window of 16 times, which corresponds to a final pixel sample of 50 cm. Once these parameters had been defined, the image corresponding to the study area was captured, resulting in a single final mosaic in geotiff format, which was used to map the land use categories for 2016.

From the aforementioned mapping, it is possible to see a predominance of herbaceous vegetation in the study area, with an occupied area percentage of 44.09%. This is followed by areas of Montane Seasonal Semideciduous Forest, with 36.17% of the area in question. Agricultural crops occupy third place, with 10.33% of the total. Next up are corridors of Alluvial Semideciduous Seasonal Forest, with 2.96% of the study area. Areas with hygrophilous vegetation, bamboo groves and groves appear next, with 0.67%, 0.02% and 0.42% of the study area, respectively. Finally, areas of cuttings, embankments and rural centres complement the land use categories for the year in question, with percentages of 0.41%, 0.19% and 0.41% respectively.

Table 2.3.a consolidates the areas obtained by the different land use classes on all the dates considered into a time frame.

The individualised observation of each period makes it possible to understand the characteristics

52

and distribution of land uses in the period considered in relation to each other, allowing us to observe the general degree of transformation of land use, such as the effects of the Rio do Peixe I and II SHPs on the landscape.

Table 2.3.a - Areas and Percentages Occupied by each land use **category** in the time periods in the study area

Land Use	1930		1972		1983		1998		2006		2016	
	ha	%	ha	%	ha	%	ha	%	ha	%	ha	%
Exposed Soil	127,63	2,35	3,13	0,05	3,26	0,06	188,20	3,48	64,33	1,18	-	-
Mowing with Exposed Soil	1,08	0,01	2,35	0,04	2,76	0,05	8,11	0,15	8,56	0,15	8,97	0,16
Cutting with Vegetation	3,54	0,06	5,36	0,09	8,12	0,15	12,43	0,22	12,81	0,23	13,72	0,25
Embankment with Exposed Soil	1,16	0,02	1,83	0,03	2,35	0,04	4,32	0,07	4,92	0,09	5,27	0,09
Landfill with Vegetation	1,37	0,02	2,36	0,04	3,21	0,05	4,59	0,08	5,13	0,09	5,65	0,10
Construction Site	3,46	0,06	-	-	-	-	-	-	-	-	-	-
Engine Room	0,03	0,0005	0,03	0,0005	0,03	0,0005	0,09	0,001	0,09	0,001	0,09	0,001
Forced Conduit	0,19	0,003	0,19	0,003	0,19	0,003	0,42	0,007	0,42	0,007	0,42	0,007
Balancing chimney	0,01	0,00018	0,01	0,00018	0,01	0,00018	0,42	0,007	0,42	0,007	0,02	0,00037
Water Intake	0,006	0,00011	0,006	0,00011	0,006	0,00011	0,01	0,00018	0,01	0,00018	0,01	0,00018
Spillway	0,07	0,001	0,07	0,001	0,07	0,001	0,11	0,002	0,11	0,002	0,11	0,002
Adduction Canal	0,35	0,006	0,35	0,006	0,35	0,006	0,53	0,009	0,53	0,009	0,53	0,009
Upstream cofferdam	0,38	0,007	-	-	-	-	-	-	-	-	-	-
Downstream cofferdam	0,49	0,009	-	-	-	-	-	-	-	-	-	-
River Diversion Area	2,14	0,03	-	-	-	-	-	-	-	-	-	-
Reservoir	31,45	0,58	31,45	0,58	31,45	0,58	80,00	1,46	80,00	1,46	80	1,46
Bodies of Water	36,19	0,66	37,45	0,69								
Unpaved road	13,48	0,24	8,14	0,15								
Transmission Line Easement Strip	-	-	3,83	0,07								
Montane Semideciduous Seasonal Forest in an Advanced Stage of Regeneration	2.275,28	42,06	547,32	10,11								
Montane Semideciduous Seasonal Forest in the Middle Stage of Regeneration	-	-	640,38	11,84								
Montane Semideciduous Seasonal Forest in the Initial Stage of Regeneration	-	-	1426,15	26,36								
Montane Semideciduous Seasonal Forest in the Pioneer Stage of Regeneration	-	-	203,36	3,76								
Alluvial Semideciduous Seasonal Forest in an Advanced Stage of Regeneration	335,85	6,20	124,36	2,29								

38,12	0,70	30,12	0,55	38,93	0,61	45,45	0,83
6,18	0,11	5,31	0,10	5,31	0,10	5,31	0,10
3,83	0,07	3,83	0,07	3,83	0,07	3,83	0,07
233,63	4,31	35,07	0,64	32,14	0,59	31,19	0,57
712,06	13,16	842,99	15,58	724,48	13,39	731,64	13,37
1026,14	18,97	1.208,83	22,35	856,26	15,83	996,98	18,22
207,48	3,83	235,56	4,35	206,12	3,81	219,50	4,01
60,15	1,11	-	-	-	-	-	-
Alluvial Semideciduous Seasonal Forest in Medium Stage of Regeneration				-	-	40,36	0,74
Stationary Alluvial Semideciduous Forest in the Initial Stage of Regeneration				-	-	96,75	1,78
Alluvial Semideciduous Seasonal Forest in Pioneer Stage of Regeneration				-	-	12,45	0,23
Herbaceous vegetation				996,75	18,42	1248,36	23,08
Hydrophilous Herbaceous Vegetation				-	-	32,25	0,59
Bamboo				-	-	0,23	0,004
Arvoredo				-	-	11,27	0,20
Reforestation				-	-	51,03	0,94
Temporary Agriculture				222,82	4,11	316,89	5,85
Permanent Agriculture				52,13	0,96	81,12	1,49
Paved road				0,54	0,009	0,96	0,01
Rural Centre				12,36	0,22	14,21	0,26
Rural Nucleus with Arvoredo				-	-	20,21	0,37

Prepared by: Edson Alves Filho

The greatest negative variations noted in land use for the study area over the entire period analysed occurred above all with the category Montane Seasonal Semideciduous Forest, considering its different successional stages: in 1925, it represented 42.06% of the study area, showing a total drop of 5.99% in 91 years. The loss noted for this class is accompanied by an increase in the areas of herbaceous vegetation (an increase of 25.67% over 91 years), temporary agriculture (an increase of 3.12% over 91 years) and permanent agriculture (an increase of 2.14% over 91 years).On the other hand, the categories of Hydrophilous Vegetation, Bamboo and Grove had small increases over the period analysed, of 0.67%, 0.02% and 0.42% respectively.When considering the area directly affected by the implementation of the Rio do Peixe I and Rio do Peixe II SHPs (study area) (Exposed Soil, Cut with Exposed Soil, Cut with Vegetation, Embankment with Exposed Soil, Embankment with Vegetation, Construction Site, Machine House, Forced Conduit, Balancing Chimney, Water Intake, Spillway,

42,52	0,78	45,86	0,84	46,12	0,85	48,28	0,88
99,36	1,83	101,40	1,87	98,14	1,81	99,08	1,81
13,18	0,24	14,67	0,27	14,62	0,27	14,63	0,27
1829,65	33,83	2.182,79	40,35	2.256,39	41,72	2.412,19	44,09
51,12	0,94	66,52	1,22	46,51	0,85	36,74	0,67
0,39	0,007	0,43	0,007	0,61	0,01	0,85	0,02
14,12	0,26	16,43	0,30	18,22	0,33	22,75	0,42
55,12	1,01	57,12	1,05	60,32	1,11	76,13	1,39
356,39	6,58	403,12	7,45	431,29	7,97	458,04	8,37

92,16	1,70	102,36	1,89	105,16	1,94	107,19	1,96
1,12	0,02	1,45	0,03	1,45	0,03	1,45	0,03
15,36	0,28	16,22	0,29	19,87	0,36	22,20	0,41
32,15	0,59	50,34	0,93	53,12	0,98	55,45	1,01

Adduction Canal, Upstream Embankment, Downstream Embankment, River Diversion Area, Reservoir, Water Bodies, Herbaceous Vegetation, Montane Semi-Deciduous Seasonal Forest in various stages of succession, Alluvial Semi-Deciduous Seasonal Forest in various stages of succession, temporary agriculture and permanent agriculture), there is little increase in the categories related to the civil infrastructure of the work, with a total increase of just 0.84%.

However, between 2006 and 2016, there was a new increase in the area occupied by vegetation cover, which interrupted the downward trend observed in the period prior to 2006. In 1998, the year immediately prior to 2006 when aerial photographs of the study area were available, the different vegetation covers accounted for 45.9 per cent of the study area, falling to 36.55 per cent in 2006.

However, in 2016, the area occupied by the vegetation cover classes rose to 39.13% of the study area, an increase of 2.58% in 10 years, justified by the change in land use in areas close to the Rio do Peixe I and II SHPP reservoirs, which
in 2006 were areas with herbaceous cover and, when left to rest, allowed the vegetation cover to recover.

However, the Environmental Impact Study for the Rio do Peixe II SHPP mentions an environmental compensation project to recover vegetation cover. However, this project was carried out with the implementation of reforestation, a fact that can be seen in the percentage occupied by this class of land use between 1998 and 2006, when there was an increase from 1.01% of this class in relation to the study area to 1.11% in 2006.

The ideal condition for assessing the effects of the implementation of the Rio do Peixe II SHPP in a situation of anthropogenic morphology during the active intervention phase would be to use aerial photographs with a time frame for the entire construction period, which in the case under study took three years (1995-1998).

The unavailability of aerial photographs for the period between 1995 and 1998, however, was not an impediment to analysing the geo-indicators of anthropogenic morphology relating to the intervention phase. The use of intermediate intervals for the physical categories of land use made it possible to understand the magnitude of the variations in landscape transformation in the study area, making it possible to note small increases between the active and consolidated intervention phases, which shows that there were no major disadvantages to not having aerial photographs for the exclusive mapping of these transformations for the years between 1995 and 1998.

As for the mapping of Geoindicators relating to Anthropo- genic Morphology in the Active Intervention Phase, in addition to the IKONOS satellite image dated 1998, the engineering project and the planialtimetric survey of the Rio do Peixe II SHPP, provided by Companhia Paulista de Força e Luz (CPFL), the current holder of the right to exploit these plants, were used.

The geo-indicators that could be spatialised were returned, making up Map 2.3.f, presented below. For those where it was not possible to delimit the spatial occurrence, the Environmental Impact Report (EIA-RIMA) for the Rio do Peixe II SHPP was used as a source of information, as was the Report on the Renewal of the Operating Licence for the Rio do Peixe II SHPP.

Among the geo-indicators systematised in Table 2.3.a, both those of a morphological/morphometric nature and those directly linked to the implementation of hydroelectric projects were applied to the condition of anthropogenic morphology in the active intervention phase.

The results obtained for the anthropogenic morphology geo-indicators related to the stage of active intervention are shown in **Table 2.3.a**:

Table 2.3.a - Geoindicators of Anthropogenic Morphology applied to the Active Intervention Phase (1998)

System	Subsystem	Indicator	Parameters	Results
Basin Hydrographic	-	Shape	Drainage pattern of the river basin	Dendritic
Basin Hydrographic	-	Shape	Surface path length of the catchment area (Eps)	23 m
Basin Hydrographic	-	Shape	Length of Channel Segments (C.s.)	113.80 kilometres
Basin Hydrographic	-	Organisation	Order of the River in the Barrage Section	4^a order
Basin Hydrographic	-	Area	Drainage Area	53.89 kilometres
Basin Hydrographic	-	Density	Hydrographic density of the catchment area (Dh)	1.84 rivers/km^2
Basin Hydrographic	-	Density	Drainage Density of the Basin Hydrographic (Dd)	2.10 km/km^2
Basin Hydrographic	-	Density	River Basin Maintenance Coefficient (Cm)	476,19 m^2
Basin Hydrographic	-	Vegetation cover	Typology	Montane Semideciduous Seasonal Forest in Advanced Stage of Regeneration (0.35 km^2), Montane Semideciduous Seasonal Forest in Medium Stage of Regeneration (8.42 km^2

				), Montane Semideciduous Seasonal Forest in Initial Stage of Regeneration (12.08 km^2), Montane Semideciduous Seasonal Forest in Pioneer Stage of Regeneration (2,35 km^2), Alluvial Semideciduous Seasonal Forest in Medium Stage of Regeneration (0.45 km^2), Alluvial Semideciduous Seasonal Forest in Initial Stage of Regeneration (1.01 km^2), Alluvial Semideciduous Seasonal Forest in Pioneer Stage of Regeneration (0.14 km)$.^2$
Basin Hydrographic	-	Vegetation cover	Successional stage	Pioneer Regeneration Stage (2.49 km^2), Initial Regeneration Stage (13.09 km^2), Medium Regeneration Stage (8.87 km^2), Advanced Regeneration Stage (0.35 km)$.^2$
Basin Hydrographic	-	Remobilised Material	Total Volumes of Works (concrete, soil, trenching, excavation)	Concrete: 150,000 m^3 , excavations (1,181,398.5 m)^3
Basin Hydrographic	-	Conditions Climate / Available Water	Average Annual Temperature	18°C
Basin Hydrographic	-	Conditions Climate / Available Water	Humidity Air Relative	50,28%
Basin Hydrographic	-	Conditions Climate / Available Water	Daily Precipitation	410 mm (December, wettest month)
Basin Hydrographic	-	Conditions Climate / Available Water	Annual rainfall	1,748 mm
Basin Hydrographic	-	Deforestation	Occupation time until new use	11 years
System Strand		Area of Occupation of Slope Classes	Area of Occupation of Slope Classes	4.97 km^2 (up to 2%), 5.23 km^2 (2% to 8%), 10.82 km^2 (8% to 15%), 23.88 km^2 (15% to 30%), 8.66 km^2 (30% to 45%), 1.14 km^2 (> 45%)
System Strand		Altitude	Average altitude of the slopes	950 m
System Strand		Guidance	Predominant Orientation	4.96 km^2 (flat), 6.16 km^2 (north), 4.22 km^2 (north-east), 5.50 km^2 (east), 6.02 km^2 (south-east), 5.39 km^2 (south), 4.46 km^2 (south-west), 8.52 km^2

				(west), 9.47 km^2 (north-west),
System Strand		Length	Average Ramp Length	2.100 m
System Strand		Soil Properties	Units of Soil	Red-Yellow Argisols (35.49 km^2), Association of Cambisols with Litholic Neosols (0.55 km^2), Association of Gleissols with Fluvic Neosols (4.06 km^2), Cambisols (9.33 km^2), Red-Yellow Latosols (4.63 km).^{2}
System Strand		Properties Soil Physics	Soil thickness	Red-Yellow Argisols (profile sizes between 120 cm and 150 cm), Association of Cambissolos with Neossolos Litholic (profiles ranging in size from 40 cm to 120 cm), Association of Gleissolos with Fluvic Neossolos (profiles ranging in size from 40 cm to 80 cm), Cambissolos (profiles ranging in size from 80 cm to 120 cm), Latossolos Vermelho-Yellow (profile sizes between 150 cm and 180 cm).
System Strand		Properties Soil Physics	Layer thickness	Argissolo Vermelho- Amarelo (Horizon A: 0-20 cm, Horizon E: 20-60 cm, Horizon Bt: 90 cm), Cambissolos (Horizon A: 0-20 cm, Horizon Bi: 20-50 cm, Horizon R: 50-70 cm), Neossolo Litólico (Horizon A: 0-20 cm, Horizon Cr: 20-30 cm, Horizon R: 30-40 cm), Gleissols (Horizon A: 0-20 cm, Horizon Ac: 20-30 cm, Horizon Cg: 30-80 cm), Fluvial Neosols (Horizon A: 0-20 cm, Horizon Bi: 20-30 cn, Horizon C: 30-80 cm), Red-Yellow Latosols (Horizon A: 0-20 cm, Horizon B: 20-70 cm, Horizon C: 70-180 cm).
System Strand		Properties Soil Physics	Structure	Red-Yellow Argisols (angular and sub-angular blocky structure), Cambisols (blocky, granular or prismatic structure), Litholic Neosols (large blocky structure), Gleissols (strong structure), Fluvic Neosols (moderate structure), Red-Yellow Latosols (small granular structure).
System		Properties	Texture	Argissolo Vermelho-Amarelo

Strand		Soil Physics		(medium/clay texture), Cambissolos (sandy loam texture), Neossolos Litólicos (sandy/sandy loam texture), Gleissolos (sandy/clay texture), Neossolo Fluvial (sandy loam/sandy loam/silty loam/silty loam), Red-Yellow Latosol (medium loam to very loamy)
System Strand		Properties Soil Physics	Porosity	Red-Yellow Argisols (49.90%), Cambisols (42.20%), Litholic Neosols (36.64%), Gleissols (34.1%), Fluvial Neosols (39.79%), Red-Yellow Latosols (51.87%).
System Strand		Disturbed Soils	Total Landfill Area	18.90 ha
System Strand		Disturbed Soils	Total Landfill Volume	58.818.241,65 m^3
System Strand		Disturbed Soils	Total volume of cuts	79.586.309,38 m^3
System Strand		Disturbed Soils	Area of Exposed Surfaces	84.30 ha
System Strand		Disturbed Soils	Material Excavated	Granites and paragneisses
System Strand		Disturbed Soils	Volume of excavations	183.716.331,79 m^3
System Strand		Lithology	Units Geological	10.95 km^2 (migmatites of the Varginha-Guaxupé Group), 16.12 km^2 (paragneisses of the Varginha-Guaxupé Group), 27.61 km^2 (charnockites of the São José do Rio Group). Pardo-Divinolândia).
System Strand		Erosion	Type of Process	Landslides, micro-terraces of trampling, mass movements, erosion processes downstream of the reservoir, slope erosion processes
System Strand		Erosion	Affected Area	72.89 ha
System Strand		Erosion	Eroded volume	252.828.482,70 m^3
System Strand		Erosion	Time of Length of Process	Average age 33.
System Strand		Erosion	Frequency	1.33 outbreaks per kilometre^2
System Strand		Deforestation	Area Cleaning the accumulation	3,08

			basin	
System Strand		Deforestation	Accumulation Basin Cleaning Time	1.3 years
River/Lake System	Plains and Terraces Subsystem	Deforestation	Accumulation Basin Cleaning Area	77.73 ha
River/Lake System	Plains and Terraces Subsystem	Deforestation	Accumulation Basin Cleaning Time	1.3 years
River/Lake System	Canal/reservoir subsystem	Area	Total Area of Water Bodies	125.44 ha
River/Lake System	Canal/ Reservoir	Perimeter	Total Perimeter of Water Bodies	60.65 kilometres
River/Lake System	Canal/ Reservoir	Area	River Plain Area	204.41 ha
River/Lake System	Canal/ Reservoir	Perimeter	Perimeter of the River Plain	51.50 km
River/Lake System	Canal/ Reservoir	Distribution	Distribution of the River Plain	It is distributed in the study area in a S - NO direction
River/Lake System	Channel/ Reservoir	Area	River Terrace Area	102.15 ha
River/Lake System	Channel/ Reservoir	Perimeter	Perimeter of River Terraces	19.11 kilometres
River/Lake System	Canal/ Reservoir	Distribution of River Terraces	-	It is divided into three areas: the first to the NW, near the confluence of the River Peixe with the River Pardo, the second to the N, located 1.4 kilometres from the first area, and finally the third, to the SO, starting from the reservoir of the Rio do Peixe I and II SHPs.
River/Lake System	Channel/ Reservoir	Disturbed Soils	Cofferdam area	1.05 ha
River/Lake System	Channel/ Reservoir	Disturbed Soils	Cofferdam volume	2.602.753,57 m3
River/Lake System	Channel/ Reservoir	Water Quality	Aluminium	N.D.
River/Lake System	Canal/ Reservoir	Water Quality	Barium	N.D.
River/Lake System	Canal/ Reservoir	Water Quality	Cadmium	N.D.
River/Lake System	Canal/ Reservoir	Water Quality	Lead	N.D.
River/Lake System	Canal/ Reservoir	Water Quality	Chloride	N.D.
River/Lake System	Canal/ Reservoir	Water Quality	Chlorophyll - a	-
River/Lake System	Canal/ Reservoir	Water Quality	Copper	N.D.
River/Lake	Canal/	Water Quality	Total Coliforms	7×10^3

River/Lake System	Canal/Reservoir	Water Quality		
River/Lak e System	Reservoir			
River/Lak e System	Canal/ Reservoir	Water Quality	Faecal coliforms	3×10^3
River/Lak e System	Canal/ Reservoir	Water Quality	Water colour	-
River/Lak e System	Canal/ Reservoir	Water Quality	Conductivity Specific	-
River/Lak e System	Canal/ Reservoir	Water Quality	Total Chromium	N.D.
River/Lak e System	Canal/ Reservoir	Water Quality	BOD	2
River/Lak e System	Canal/ Reservoir	Water Quality	COD	6
River/Lak e System	Canal/ Reservoir	Water Quality	Phenols	N.D.
River/Lak e System	Canal/ Reservoir	Water Quality	Total Iron	N.D.
River/Lak e System	Channel/ Reservoir	Water Quality	Total Phosphorus	0,255
River/Lak e System	Canal/ Reservoir	Water Quality	Manganese	N.D.
River/Lak e System	Canal/ Reservoir	Water Quality	Mercury	N.D.
River/Lak e System	Canal/ Reservoir	Water Quality	Microtox	-
River/Lak e System	Channel/ Reservoir	Water Quality	Nickel	N.D.
River/Lak e System	Channel/ Reservoir	Water Quality	Ammoniacal Nitrogen	0,11
River/Lak e System	Channel/ Reservoir	Water Quality	Nitrogen Nitrate	0,64
River/Lak e System	Canal/ Reservoir	Water Quality	Nitrogen Nitrite	0,01
River/Lak e System	Canal/ Reservoir	Water Quality	Total Kjedahl Nitrogen	2,0
River/Lak e System	Canal/ Reservoir	Water Quality	Oils and greases	N.D.
River/Lak e System	Canal/ Reservoir	Water Quality	Soluble Orthophosphate	-
River/Lak e System	Canal/ Reservoir	Water Quality	Dissolved Oxygen (DO)	7,5
River/Lak e System	Canal/ Reservoir	Water Quality	pH	6,7
River/Lak e System	Canal/ Reservoir	Water Quality	Non-Filterable Waste	28
River/Lak e System	Canal/ Reservoir	Water Quality	Total Waste	74
River/Lak e System	Canal/ Reservoir	Water Quality	Solids Suspended Totals	-
River/Lak e System	Canal/ Reservoir	Water Quality	Surfactants	N.D.

River/Lake System	Canal/Reservoir	Water Quality	Test Toxicity Chronicle	N.D.
River/Lake System	Canal/Reservoir	Water Quality	Turbidity	-
River/Lake System	Canal/Reservoir	Water Quality	Zinc	N.D.
River/Lake System	Canal/Reservoir	Flow rate	Main Channel Flow (Q7, Q10)	11.75 m^3 /s (average) Q7.10 (1.03 m^3 /s), Long Term Average Flow (7.26 m^3 /s)
River/Lake System	Canal/Reservoir	Flow rate	Peak Flow	14.61 m^3 /s
River/Lake System	Canal/Reservoir	Flow rate	Existence of a Bottom Discharger	Yes
River/Lake System	Canal/Reservoir	Dynamics of Sedimentation	Sedimentary Supply	3.26 tonnes/km2/year
River/Lake System	Canal/Reservoir	Flooding	Event Location Point	Rio do Peixe II SHPP Fluviometric Station (21° 36' 43" S, 46° 47'37"W
River/Lake System	Canal/Reservoir	Flooding	Number of Events	12 records between 1953-1987
River/Lake System	Canal/Reservoir	Flooding	N.A. Achieved	2,9 m
River/Lake System	Canal/Reservoir	Flooding	Maximum flows during events	34.6 m^3 /s
River/Lake System	Canal/Reservoir	Flooding	Duration of events	11 hours
River/Lake System	Canal/Reservoir	Time	Time of Filling the Reservoir	2.4 months
River/Lake System	Channel/Reservoir	Speed	Reservoir Filling Speed	639,360 m /day^3
River/Lake System	Channel/Reservoir	Length	Length of the Reduced Flow Section (between the dam and the powerhouse)	853,97 m
River/Lake System	Canal/Reservoir	Length	Extension of River diversion	264 m
River/Lake System	Canal/Reservoir	Perimeter	Reservoir perimeter	8.42 kilometres
River/Lake System	Canal/Reservoir	Width	Channel Section Width (Wmp)	22,50 m
River/Lake System	Channel/Reservoir	Depth	Depth Section Average (Dmp)	2,20 m
River/Lake System	Channel/Reservoir	Volume	Volume Total Reservoir	60.000.000 m^3
River/Lake System	Channel/Reservoir	Slope	Profile Longitudinal Channel	Characteristic equilibrium profile with knick points at the confluence of higher order

				channels.
River/Lak e System	Canal/ Reservoir	Slope	River Gradient	24,52 m
River/Lak e System	Canal/ Reservoir	Cross-section	Cross-sectional area (Amp)	19,31 m^2
River/Lake System	Canal/ Reservoir	Cross-section	Wet area (A)	19,28 m^2
River/Lake System	Canal/ Reservoir	Cross-section	Perimeter Wet (P)	26,90 m
River/Lake System	Canal/ Reservoir	Cross-section	Lightning Hydraulic (R)	0.71 m/m^2
River/Lake System	Canal/ Reservoir	Shape and Pattern of the Watercourse	Channel Shape Index (F)	0,25
River/Lake System	Canal/ Reservoir	Shape and Pattern of the Watercourse	Sinuosity Index (Is)	2,50

Prepared by: Edson Alves Filho

The mapping of land use categories for 1998 provided important evidence of the influence of the Rio do Peixe II SHPP on the geo-morphological systems impacted in the study area. Consultation of the engineering project and the Environmental Impact Study for the Rio do Peixe II SHPP provided access to information on the sizing of this hydroelectric project, as well as various environmental indices used in the environmental licensing process.

For the geo-indicators used for the catchment system, the comparison between the condition of anthropogenic morphology in the **active intervention phase** and the condition of preserved original morphology in the pre-intervention condition revealed a picture of significant changes.

Although the *drainage pattern of the* hydrographic basin and *the river order* in the stretch of the dam did not show major changes between the phases being compared, as is the case with the region near the Rio do Peixe II SHPP dam, this is not the case with the other indicators relating to the hydrographic basin system. For the indicators *Extension of the Surface Course of the Hydrographic Basin Length of River Channel Segments, Hydrographic Density of the Hydrographic Basin, Drainage Den- age of the Hydrographic Basin* and Coe/zczewte *of* Afawwtewtion, the changes in the comparison of semi-preserved original morphology and anthropogenic morphology (phase of active intervention) were respectively 84.6 per cent (reduction from 150 m to 123 m), 33.6 per cent (reduction from 171.56 km to 113.80 km), 60 per cent (reduction from 4.60 rivers/km^2 to 1.84 rivers/km^2), 32.9 per cent (reduction from 3.13 km/km^2 to 2.10 km/km^2) and an increase from 310 m^2 /m to 476.19 m^2 /m (an increase of 34.8 per cent).

For the climatic condition indicators, due to the existence of time projections in the

Environmental Impact Study, it was possible to compare the situation between the conditions of semi-preserved original morphology (pre-intervention phase) and anthropogenic morphology (active intervention phase). The indicators *annual average temperature, relative humidity, daily rainfall* and *annual precipitation* showed variations of 4.4 per cent (increase from 17.2°C to 18°C), 3.8 per cent (reduction from 52.3 per cent to 50.28 per cent), 32.5 per cent (increase from 8.91 mm to 13.2 mm) and 31.2 per cent (increase from 1,201.9 mm to 1,748 mm) respectively.

These increases, especially in relation to the daily and annual rainfall variable, can be attributed, as explained in the work by NASCIMENTO & NERY (2005), to the *El Nino* phenomenon in the north-eastern part of the state of São Paulo, with a considerable increase in rainfall being observed between the climatological normals from 1941 to 1970 and in the 1971 - 2000 climatological normal, due to the little influence of *El Nino in* this second period. The considerable increase in atmospheric disturbance, coupled with the orographic effect in the São José do Rio Pardo region, may have been factors that further boosted the increase in rainfall in the 1971 - 2000 climatological normal period, thus justifying the sharp increase in daily and annual rainfall in the study area when comparing the situations of semi-preserved original morphology in the pre-intervention phase and anthropogenic morphology in the active intervention and post-intervention phases.

When analysing the indicators listed for the slope system, more precisely the *deforestation* promoted by the implementation of the project, the mapping of land use categories and associated morphological interventions was used to verify the change in use between the two conditions in question. A decrease of 2% over 73 years can be seen for *Modtana Semideciduous Stationary Forest* (reduction in the total area occupied from 42.92% in 1925 to 42.06% in 1998) and *yá/wv/a/ Semideciduous Stationary Forest, with* a percentage of 3.22% (reduction from 6.20% of the total area to 2.98% of the total area) over the 73 years analysed.

With regard to the Jec/zvzJatZe *classes,* using the contour lines from the Casa Branca Sheet made it possible to draw up an MDT and calculate the slope classes in the condition of semi-preserved original morphology. For the anthropogenic morphology condition, the slope map drawn up for the study area was used (Figure 4.3.b, Volume 1). The comparison between the two maps showed a decrease in the slope range of up to 2% (from 5.20 km^2 to 4.97 km^2), a decrease in the slope range between 2% and 8% (from 5.32 km^2 to 5.23 km^2), an increase in the slope classes between 8% and 15% (10,26 km^2 to 10.82 km^2), an increase in the slope class between 15% and 30% (23.42 km^2 to 23.88 km^2), an increase in the slope class between 30% and 45% (7.42 km^2 to 8.66 km^2) and, finally, an increase in the slope class greater than 45% (1.02 km^2 to 1.14 km).^{2}

With regard to the indicators *average altitude, predominant orientation and average ramp length, there* was a 1.05% reduction for the former (950 m in the semi-preserved original morphology condition to 940 m in the anthropogenic morphology condition in the active intervention phase), maintenance of the northerly direction between the two phases analysed and a 0.85% reduction in ramp length between the two conditions analysed (reduction from 2,100 m in the semi-preserved original morphology condition to 2,082 m in the anthropogenic morphology condition in the active intervention phase).100 m in the semi-preserved original morphology condition in the pre-intervention phase to 2,082 m in the anthropogenic morphology condition in the active intervention phase).

The indicators listed for assessing the soil's physical properties were obtained from the tests carried out for the Executive Engineering Project, and it is not possible to compare the data found in relation to the situation of semi-preserved original morphology. The indicators will therefore be presented in a watertight manner, with no comparative perspective.

The soil units found along the slope system are red-yellow argisols, *cambisols* and verzweZÃo-awareZo *latosols, with* average thicknesses of 1.50 m for the standard red-yellow argisol profiles and horizons with average thicknesses of 30 cm, standard profiles of cambisols of around 1 m, with average thicknesses between horizons of 35 cm, and finally standard profiles of red-yellow latosols of 1.80 m, with average profile thicknesses of 45 cm. The structure of argisols is generally angular and sub-angular blocks, while cambisols are blocks, grains or prisms, and finally, latosols are small grains. The predominant texture of the argisols is medium to loamy, of the cambisols is loamy and of the latosols medium loamy to very loamy.

Comparison with the condition of the original semi-preserved morphology is not possible in relation to the *parturbated soil* indicators, since these indicators are typical of anthropogenic actions (cuts and embankments), which are only noticeable during the active and consolidated intervention phases. We found a total area and volume of embankments of around 18.90 ha and 58,818,241.65 m^3 , respectively, distributed along the motorway.
SP-211, which crosses the study area. For the volume of cuts, an area of 84.30 ha was found, which corresponds to a volume of 79,586,309.38 m^3 , spread throughout the study area, but following the existing roads.

With regard to the material excavated and the volume of excavations, the information obtained from the Environmental Impact Study (EIA-RIMA) for the Rio do Peixe II SHPP shows that a volume of 183,716,331.79 metres^3 of granite and paragneiss was excavated to build the civil structures for the Rio do Peixe II SHPP.

When continuing the analysis of the indicators for the *current system, it* is worth noting that there were no geological surveys for the pre-intervention phase, so the mapping of lithostratigraphic units was used (Map 6.2.a, Volume I). As a solution, the quantities from this mapping were used to characterise both the pre-intervention phase and the active intervention phase, thus repeating the values mentioned above (**Section 2.2**).

With regard to erosion processes, a total area of 72.89 ha was found, which represents an eroded volume of 252,828,482.70 m^3 . The frequency of the mapped processes was estimated at 33 years between the occurrence of an erosion focus in the condition of semi-preserved original morphology and the stabilisation or end of the process during the active intervention phase.

For the deforestation indicators applied to the plains and terraces subsystem, the source of information used was the Environmental Impact Study (EIA -RIMA) for the Rio do Peixe II SHPP, which found an accumulation basin area of 77.73 km^2 . The time taken to clean the basin was 1.3 months.

The area and *perimeter of the fluvial* jz/awzcze found were 204.41 ha and 60.65 km respectively, distributed in a NW-SE direction. As for the river terraces, the area and perimeters found were 102.15 ha and 19.11 km, distributed to the NW, N and SO of the study area.

When analysing the indicators selected for the *channel/reservoir* subsystem, *the* area and *wetted* area of *the river channel were* 19.28 m^2 and 26.90 m respectively. With regard to the *partuebated soils* indicator, the area and volume of the cofferdams were taken from the Rio do Peixe II SHPP Environmental Impact Study, which found a total area of 1.05 ha and a volume of 2,602,753.57 m^3 .

Still with regard to the indicators for the canal/reservoir subsystem, it is worth noting that 36 parameters were analysed for the *water quality* indicator, the data for which is included in the Environmental Impact Study (EIA-RIMA) for the Rio do Peixe II SHPP. Of all the parameters analysed, it is worth noting that the total coliform, faecal coliform, total phosphorus and nitrate nitrogen parameters showed values above what is tolerable according to the parameters defined by CETESB and IBAMA, with percentages above what is recommended in 20.3%, 32.8%, 2.7% and 3.6%.

With regard to the hydrodynamic indicators listed for the river channel/reservoir subsystem, it should be noted that the values indicated for the *flow of the Pomclpae channel* were 11.75 m^3 /s for the *average flow,* 1.03 m^3 /s for the *Q7.10 Flow* and 7.26 m^3 /s for the *Long Term Average Flow.*

The hydrological studies consulted revealed a total of 12 zMzzzation *episodes* in the historical series analysed, measured by the flow data from the Rio do Peixe II SHPP fluvial metric station. The *average height* reached by these events was 2.9 metres, caused by flows exceeding 34.6 metres^3 /s.

Consultation of the Environmental Impact Study for the Rio do Peixe II SHPP revealed an *average reservoir filling* time of 2.4 months, with a calculated velocity of 639,360 m^3 /day. The length of the reduced flow stretch was indicated as 853.97 metres.

In order to install the upstream and downstream cofferdams, it was necessary to make a diversion in the River Peixe of around 264 metres.

As for the *geometry/wvzaZ* indicators, the *perimeter of* the *reservoir is around* 8.42 km. The area of the reservoir *after filling* was calculated at 95.58 ha and its *total* voZz/zne at 60,000,000 m^3 . The *width of the main channel section, as* well as the *average depth* and area were calculated at 22.50 m, 2.20 m and 19.31 m² respectively.

The *longitudinal profile of the river channel* was classified as an equilibrium profile, with *knickpoints* at the confluence of the higher-order channels. The *Fluvial Gradient* found was 24.52 metres.

When analysing the indicators relating to *Form*, it was possible to determine a Channel Form Index of 0.25.

In the regions around the reservoir and on the slopes close to it, it was possible to see the biggest changes that the implementation of the Rio do Peixe I and II SHPs caused in the landscape, with emphasis on the volume of silted areas, the triggering of erosion processes and the increase in the number of floods.

When comparing the situation of semi-preserved original morphology in the pre-intervention phase with the situation of anthropogenic morphology in the active disturbance phase, there are obvious changes in the slope system in terms of the indicators increase in *eroded surface, decrease in slope gradient* and *suppression of original land cover. AS FOR* the plain and terrace subsystem, there is an increase in the *surface area of cuts* and *embankments*.

In order to specialise the geo-indicators in the condition of Anthropogenic Morphology in the Phase of Active Intervention, **Map 2.3.f** - Anthropogenic Morphology in the Phase of Active Intervention in the Study Area, presented below, was drawn up.

With regard to the point elements of representation, the tubular wells registered by the São Paulo State Department of Water and Electricity (DAEE) were located, as were the septic tanks on the rural properties in the study area, which were confirmed during fieldwork in the study area with the residents of the ranches and farms.

As linear elements of representation, the existing paths in the coffee, sugar cane, corn and vegetable plantations of the study area were mapped because of their ability to change the dynamics of surface runoff in the area, such as promoting erosion processes. Next, the areas of the dam with cracks were mapped, a phenomenon detected during fieldwork with GPS point recording and confirmed by reading the Rio do Peixe II SHPP Operation Reports, which appear as annexes to the Rio do Peixe II SHPP Operation Licence Renewal Report.

This report shows photographs of the areas of the dam with cracks and indicates treatment therapies to solve the problem. The areas of rectified hydrography were noted when comparing the situation of semi-preserved original morphology in the pre-intervention phase with the situation of anthropogenic morphology in the active intervention phase, and the same procedure was followed for buffered hydrography. Finally, mass movements and furrows were mapped by visual interpretation of the 1998 IKONOS satellite image.

With regard to the areal elements of representation, mention should be made of rural agglomerations, areas of temporary agriculture, areas of permanent agriculture, abandoned mining pits, reservoir backwater deposits, areas with rockslides, areas with a distribution of macrophytes, areas with the presence of landslide scars, roads, areas with the formation of deposits at river mouths, transmission lines, areas with trampling microterraces, areas with mass movement scars, areas triggering erosion processes upstream and downstream of the Rio do Peixe II SHPP reservoir, the Rio do Peixe II SHPP reservoir area, areas with exposed soil and earth movements and collapsed areas, mapped by visual interpretation of these features on the IKONOS satellite image mentioned above.

The landfill areas of terraces and river plains were taken from the Basic Project for the Rio do Peixe II SHPP, where there is an indication of the work's landfill skirts, including those located within the aforementioned geomorphological compartments. The areas of tectogenic deposits, the civil structures of the Rio do Peixe I and II SHPs, borrow areas and subsidence areas due to the opening of tunnels and canals were taken from the Rio do Peixe II SHP Basic Project, where there are plans in dwg format showing these structures.

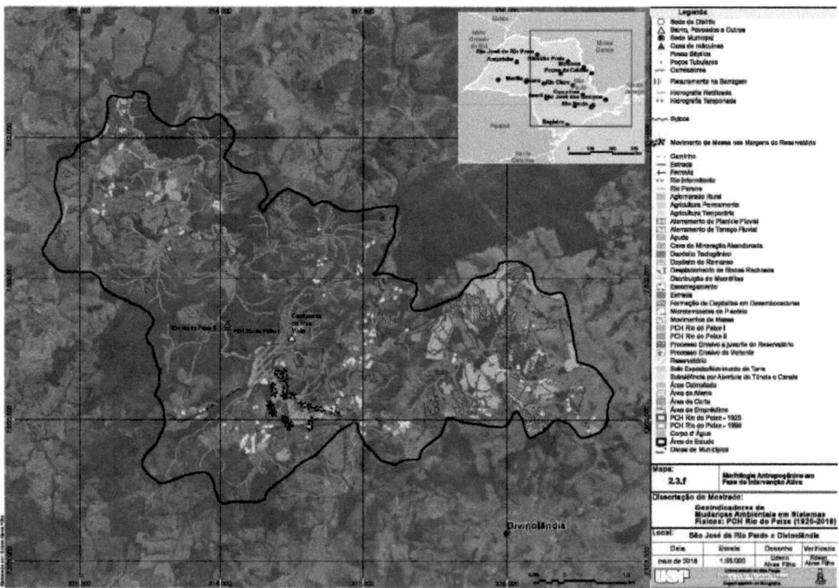

Map 2.3.f - Anthropogenic Morphology in the Phase of Active Intervention in the Study Area

Map 2.3.f shows that the areas with septic tanks are concentrated in the centre and south of the study area, following the agglomerations. rural areas, while the areas with tube wells are concentrated in the eastern part, next to an area with a strong presence of horticulture and temporary crops, which are quite intensive in terms of water use.

The areas with plantation tracks are also concentrated in the eastern part of the study area, where agricultural use predominates. The dam areas with cracks are located to the left of the Rio do Peixe II SHPP bottom spillway, representing natural wear and tear of the dam structures due to hydroelectric use.

The areas with rectified hydrography are located in the south-west of the study area and represent rectifications in tributaries of the Cachoeira stream. The areas with buffered hydrography are located in the south-west and north-west of the study area, following the route of the SP-212 motorway, as a series of small pinguelas installed over the tributaries of the streams in the study area, allowing people to cross over the existing side roads.

On the other hand, the areas with mass movements on the reservoir bank are concentrated in the region near the Rio do Peixe II SHPP dam, and finally, the areas with the presence of furrows are scattered throughout the study area, with a slightly higher concentration in the central and eastern portions, where agricultural use is more important.

With regard to the areal elements considered in the aforementioned map, there is a greater concentration of rural settlements in the south-western, central and southern portions of the study area. Temporary and permanent agricultural areas are concentrated to the north and east, in the upper and middle portions of the slopes.

The areas with embankments and river plains are located further south, along the route of State Highway SP-212. The weir areas follow the distribution of rural settlements and are more numerous in the east, centre and south of the study area.

The areas with abandoned mining pits are located in the immediate backwater of the Rio do Peixe II SHPP reservoir, on river terraces. The areas of tectogenic deposits accompany the route of State Highway SP-212 and the region close to the Rio do Peixe II SHPP dam, and therefore reflect the materials that were remobilised during the installation of these structures.

The areas with backwater deposits in the Rio do Peixe II SHPP reservoir are concentrated in three positions: a first near the dam, a second in the central position of the reservoir, and a third, of greater spatial expression, in the outermost backwater position, near the abandoned mining pits in the study area.

The areas with rock blocks are located in three main clusters in the study area, a first to the north-west, in the headwaters region near the Moinho stream, a second, in the central portion, near the Rio do Peixe II SHPP dam, and a third, to the east, in the headwaters region of the Boa Vista stream.

The areas with macrophytes are located in the central portion of the reservoir, close to the Rio do Peixe II SHPP dam. The areas with landslide scars are located in the north-west and south-west of the study area, in the middle third of the slopes, surrounded by furrows along the headwaters of the Moinho and Cachoeira streams.

As for the roads, they are widespread throughout the study area, with a slight concentration in the region around the dam and reservoir of the Rio do Peixe II SHPP. The mouth deposits are located in the tributaries that cross Fazenda Santa Amélia and flow into the Rio do Peixe and in the tributaries that cross Fazenda Ouro Branco and flow into the Rio do Peixe.

The easements of the transmission lines that cross the study area are located in the northern part of the study area, near the confluence of the Peixe River with the Pardo River. The areas with microterraces of trampling are grouped into three strips in the study area: a first, to the northwest, within the boundaries of Fazenda Santa Isabel, a second, in the centre, within the boundaries of Fazenda Bela Vista, and finally, an eastern one, within the boundaries of Fazendas Boa Vista and Santa Virgínia.

There were only two regions with mass movement scars in the study area, one in the centre, near the Boa Vista waterfall, and the other to the east, in a region of dividers within the boundaries of the Samambaia site.

The areas with erosion processes downstream of the Rio do Peixe II SHPP dam are concentrated along the course of the River Peixe in front of the borders with the Ouro Branco, Santa Isabel and Rio do Peixe farms. The upstream erosion areas are located in the watercourses to the south-west and east of the study area, more precisely in the watercourses located within the Santa Amélia and Boa Vista farms.

The areas with exposed soil and earth movement are located in the south-western part of the study area, within areas located in the upper part of the slopes and abandoned in relation to agricultural use.

The areas of subsidence due to tunnelling and plugging are located within the civil structures of the Rio do Peixe I and II HPPs, more precisely in the central part of the study area. The areas with subsidence due to tunnelling are located immediately outside the engine room of the Rio do Peixe I and II hydroelectric power plants, where the penstocks of the two plants pass through, and the filled-in areas are located in the outermost position of the backwater, in the southern portion of the reservoir, close to the border with Fazenda Santa Amélia.

Finally, the cuttings, embankments and borrow areas, which represent materials that were remobilised during the implementation of the Rio do Peixe I and II SHPPs, the SP-212 State Highway and the side roads, are scattered throughout the study area, but are concentrated in the central, south-western and eastern portions, precisely where the largest farms in the study area and in the region where the Rio do Peixe I and II SHPPs were implemented are located.

2.4 Application of Geoindicators related to Anthropogenic Morphology (Phase Consolidated Intervention)

The geo-indicators selected for the **consolidated intervention phase** showed similar results to those applied in the active intervention phase, even when taking into account the 18-year period between the two phases.

Table 2.4.a shows the results found for the anthropogenic morphology geo-indicators in the consolidated intervention phase.

Table 2.4.a - Geoindicators of Anthropogenic Morphology applied to the Consolidated Intervention Phase (2016)

System	Subsystem	Indicator	Parameters	Results
Basin Hydrographic	-	Shape	Drainage Pattern of the River Basin	Dendritic
Basin Hydrographic	-	Shape	Surface path length of the catchment area (Eps)	23 m
Basin Hydrographic	-	Shape	Length of Channel Segments (C.s.)	113.80 kilometres
Basin Hydrographic	-	Area	Drainage Area	53.89 kilometres^2
Basin Hydrographic	-	Density	Hydrographic density of the catchment area (Dh)	1.84 rivers/km^2
Basin Hydrographic	-	Density	Drainage Density of the Basin Hydrographic (Dd)	2.10 km/km^2
Basin Hydrographic	-	Density	River Basin Maintenance Coefficient (Cm)	476,19 m^2
Basin Hydrographic	-	Vegetation cover	Typology	Montane Semideciduous Seasonal Forest in the Advanced Stage of Regeneration (0.57%), Montane Semideciduous Seasonal Forest in the Middle Stage of Regeneration (13.37%), Montane Semideciduous Seasonal Forest in the Initial Stage of Regeneration (18.22%), Montane Seasonal Forest in the Pioneer Stage of Regeneration (4.01%), Alluvial Semideciduous Seasonal Forest in Advanced Stage of Regeneration (0.0001%), Alluvial Semideciduous Seasonal Forest in Medium Stage of Regeneration (0.88%), Alluvial Semideciduous Seasonal Forest in Initial Stage of Regeneration (1.81%), Alluvial Seasonal Forest in Pioneer Stage of Regeneration (0.27%)

Basin Hydrographic	-	Vegetation cover	Successional stage	Advanced Stage of Regeneration (31.19 ha), Middle Stage of Regeneration Stage (61.65 ha), Initial Regeneration Stage (1,096.06 ha), Pioneer Regeneration Stage (234.13 ha)
Basin Hydrographic	-	Conditions Climate / Available Water	Temperature Annual Average	20,5°C
Basin Hydrographic	-	Conditions Climate / Available Water	Humidity Air Relative	66,51%
Basin Hydrographic	-	Conditions Climate / Available Water	Daily Precipitation	437.5 mm (January, wettest month)
Basin Hydrographic	-	Conditions Climate / Available Water	Precipitation Annual	1,596.2 mm
System Strand	-	Slopes	Area of Occupation of slope classes	4.97 km^2 (up to 2%), 5.23 km^2 (2% to 8%), 10.82 km^2 (8% to 15%), 23.88 km^2 (15% to 30%), 8.66 km^2 (30% to 45%), 1.14 km^2 (> 45%)
System Strand	-	Altitude	Average altitude of the slopes	'950 m
System Strand	-	Guidance	Guidance Predominant	4.96 km^2 (flat), 6.16 km^2 (north), 4.22 km^2 (north-east), 5.50 km^2 (east), 6.02 km^2 (south-east), 5.39 km^2 (south), 4.46 km^2 (south-west), 8.52 km^2 (west), 9.47 km^2 (north-west),
System Strand	-	Length	Length Mid Ramp	2.100 m
System Strand	-	Processes Strand	Erosive Processes, Mass Movements	72.89 ha
System Strand	-	Disturbed Soils	Total area of Landfill sites	18.90 ha
System Strand	-	Disturbed Soils	Total Volume Landfill sites	58.818.241,65 m3
System Strand	-	Disturbed Soils	Cutting area	226.906,15 m2
System Strand	-	Disturbed Soils	Total volume of cuts	79.586.309,38 m3
System Strand	-	Disturbed Soils	Area of Exposed Surfaces	84.30 ha
System Strand	-	Lithology	Units Geological	10.95 km^2 (migmatites of the Varginha-Guaxupé Group), 16.12 km^2 (paragneisses of the Varginha-Guaxupé Group), 27.61 km^2 (charnockites of the São José do Rio Group). Pardo-Divinolândia).
System Strand	-	Erosion	Type of Process	Landslides, micro-terraces of trampling, mass movements, erosion processes downstream

				of the reservoir, slope erosion processes
System Strand	-	Erosion	Affected Area	72.89 ha
System Strand	-	Erosion	Eroded volume	252.828.482,70 m^3
System Strand	-	Erosion	Time of Length of Process	Average age 33.
System Strand	-	Erosion	Frequency	1.33 outbreaks per kilometre^2
River/Lake System	Plains and Terraces Subsystem	Altitude	Maximum Flood Elevation	682,9 m
River/Lake System	Plains and Terraces Subsystem	Area	Maximum Flood Area	265.78 ha
River/Lake System	Plains and Terraces Subsystem	Area	Floodplain area suppressed by the reservoir	12.88 ha
River/Lake System	Plains and Terraces Subsystem	Reforestation	APP Revegetation Time	6 years
River/Lake System	Plains and Terraces Subsystem	Reforestation	Area Reforestation of the APP	50 ha
River/Lake System	Plains and Terraces Subsystem	Area	Total Area of Water Bodies	125.44 ha
River/Lake System	Plains and Terraces Subsystem	Perimeter	Total Perimeter of Water Bodies	60.65 kilometres
River/Lake System	Plains and Terraces Subsystem	Area	River Plain Area	204.41 ha
River/Lake System	Plains and Terraces Subsystem	Perimeter	Perimeter River Plain	51.50 km
River/Lake System	Plains and Terraces Subsystem	Distribution	Distribution of the River Plain	It is distributed in the study area in the S-NO direction
River/Lake System	Plains and Terraces Subsystem	Area	River Terrace Area	102.15 ha
River/Lake System	Plains and Terraces Subsystem	Perimeter	Perimeter of River Terraces	19.11 kilometres
River/Lake System	Plains and Terraces Subsystem	Distribution	Distribution of River Terraces	It is divided into three areas: the first to the NW, near the confluence of the River Peixe with the River Pardo, the second to the N, located 1.4 kilometres from the first area, and finally the third, to the SO, starting from the reservoir of the Rio do Peixe I and II SHPs.
River/Lake	Canal/reservo	Water Quality	Aluminium	-

River/Lak e System	Canal/reservo ir subsystem			
River/Lak e System	Canal/reservo ir subsystem	Water Quality	Barium	-
River/Lak e System	Canal/reservo ir subsystem	Water Quality	Cadmium	0,05
River/Lak e System	Canal/reservo ir subsystem	Water Quality	Lead	0,05
River/Lak e System	Canal/reservo ir subsystem	Water Quality	Chloride	-
River/Lak e System	Canal/reservo ir subsystem	Water Quality	Chlorophyll - a	0,5
River/Lak e System	Canal/reservo ir subsystem	Water Quality	Copper	0,05
River/Lak e System	Canal/reservo ir subsystem	Water Quality	Total Coliforms	17.000
River/Lak e System	Canal/reservo ir subsystem	Water Quality	Faecal coliforms	200
River/Lak e System	Canal/reservo ir subsystem	Water Quality	Water colour	-
River/Lak e System	Canal/reservo ir subsystem	Water Quality	Conductivity Specific	56,2
River/Lak e System	Canal/reservo ir subsystem	Water Quality	Total Chromium	0,01
River/Lak e System	Canal/reservo ir subsystem	Water Quality	BOD	1
River/Lak e System	Canal/reservo ir subsystem	Water Quality	COD	229
River/Lak e System	Canal/reservo ir subsystem	Water Quality	Phenols	1
River/Lak e System	Canal/reservo ir subsystem	Water Quality	Total Iron	-
River/Lak e System	Canal/reservo ir subsystem	Water Quality	Total Phosphorus	0,09
River/Lak e System	Canal/reservo ir subsystem	Water Quality	Manganese	-
River/Lak e System	Canal/reservo ir subsystem	Water Quality	Mercury	-
River/Lak e System	Canal/reservo ir subsystem	Water Quality	Microtox	-
River/Lak e System	Canal/reservo ir subsystem	Water Quality	Nickel	-
River/Lak e System	Canal/reservo ir subsystem	Water Quality	Ammoniacal Nitrogen	0,04
River/Lak e System	Canal/reservo ir subsystem	Water Quality	Nitrogen Nitrate	1,5
River/Lak e System	Canal/reservo ir subsystem	Water Quality	Nitrogen Nitrite	0,015
River/Lak e System	Canal/reservo ir subsystem	Water Quality	Total Kjedahl Nitrogen	0,72
River/Lak e System	Canal/reservo ir subsystem	Water Quality	Oils and greases	0,1
River/Lak e System	Canal/reservo ir subsystem	Water Quality	Soluble Orthophosphate	-
River/Lak e System	Canal/reservo ir subsystem	Water Quality	Oxygen Dissolved (DO)	7,00

River/Lake System	Canal/reservoir subsystem	Water Quality	pH	7,4
River/Lake System	Canal/reservoir subsystem	Water Quality	Non-Filterable Waste	-
River/Lake System	Canal/reservoir subsystem	Water Quality	Total Waste	-
River/Lake System	Canal/reservoir subsystem	Water Quality	Solids Total Suspense	73
River/Lake System	Canal/reservoir subsystem	Water Quality	Sulphating agents	-
River/Lake System	Canal/reservoir subsystem	Water Quality	Chronic Toxicity Test	-
River/Lake System	Canal/reservoir subsystem	Water Quality	Turbidity	18
River/Lake System	Canal/reservoir subsystem	Water Quality	Zinc	0,05
River/Lake System	Canal/reservoir subsystem	Flow rate	Main Channel Flow (Q7, Q10)	Long-term average flow (7.4 m^3/s), Q7.10 flow (1.63 m^3/s)
River/Lake System	Canal/reservoir subsystem	Flow rate	Peak Flow	14.61 m^3/s
River/Lake System	Canal/reservoir subsystem	Flow rate	Existence of a Bottom Discharger	Yes
River/Lake System	Canal/reservoir subsystem	Flow rate	Frequency of bottom unloader opening	Whenever the flow exceeds 14.32 m^3/s
River/Lake System	Canal/reservoir subsystem	Flow rate	Flow during bottom spillway opening procedure	32 m^3/s (based on maximum monthly flows)
River/Lake System	Canal/reservoir subsystem	Change in N.A.	Reservoir Operating Regime	Water thread
River/Lake System	Canal/reservoir subsystem	Change in N.A.	Daily reservoir depletion	Measurement on 21/08/2007: Maximum N.A.: 860.50; Minimum N.A.: 860.33 m. Variation: 17 cm
River/Lake System	Canal/reservoir subsystem	Change in N.A.	Monthly reservoir depletion	-
River/Lake System	Canal/reservoir subsystem	Change in N.A.	Annual reservoir depletion	Measurement in June 2006: N.A. 859.50 m (1.50 m below operating level). Measurement in Aug/2007: N.A.: 860.33 m (67 cm below operating level). Level Minimum operational: 861 m
River/Lake System	Canal/reservoir subsystem	Dynamics of Sedimentation	De-silting rate	Reservoir silted volume in 1998: 0%; silted volume in 2006: 8.8%
River/Lake System	Canal/reservoir subsystem	Flooding	Event Location Point	0
River/Lake System	Canal/reservoir subsystem	Flooding	Number of Events	0
River/Lake System	Canal/reservoir subsystem	Flooding	N.A. Achieved	-
River/Lake System	Canal/reservoir subsystem	Flooding	Maximum flows during events	Maximum flow reached in 2016: 14.61 m^3/s, close to the opening flow of the bottom

				spillway: 14.32 m³ /s
River/Lak e System	Canal/reservo ir subsystem	Flooding	Duration of events	-
River/Lak e System	Canal/reservo ir subsystem	Length	Length of the Reduced Flow Section (between the dam and the powerhouse)	853,97 m
River/Lak e System	Canal/reservo ir subsystem	Perimeter	Reservoir perimeter	8.42 kilometres
River/Lak e System	Canal/reservo ir subsystem	Width	Channel Section Width (Wmp)	22,50 m
River/Lak e System	Canal/reservo ir subsystem	Width	Width Surface (L)	12,25 m
River/Lak e System	Canal/reservo ir subsystem	Depth	Average Section Depth (Dmp)	2,20 m
River/Lak e System	Canal/reservo ir subsystem	Volume	Total Reservoir Volume	60.000.000 m³
River/Lak e System	Canal/reservo ir subsystem	Volume	Useful Volume Reservoir	47.500.000 m³
River/Lak e System	Canal/reservo ir subsystem	Slope	Profile Longitudinal Channel	Characteristic equilibrium profile with knick points at the confluence of higher order channels.
River/Lak e System	Canal/reservo ir subsystem	Slope	River Gradient	24,52 m
River/Lak e System	Canal/reservo ir subsystem	Cross-section	Cross-sectional area (Amp)	19,31 m²
River/Lak e System	Canal/reservo ir subsystem	Cross-section	Wet area (A)	19,28 m²
River/Lak e System	Canal/reservo ir subsystem	Cross-section	Perimeter Wet (P)	26,90 m
River/Lak e System	Canal/reservo ir subsystem	Cross-section	Hydraulic Radius (R)	0.71 m/m²
River/Lak e System	Canal/reservo ir subsystem	Shape and Pattern of the Watercourse	Extension Suppressed by channel shape	4.12 km of meandering watercourse
River/Lak e System	Canal/reservo ir subsystem	Shape and Pattern of the Watercourse	Channel Shape Index (F)	0,25
River/Lake System	Canal/reservo ir subsystem	Shape and Pattern of the Watercourse	Original bed sinuosity index	2,50

Prepared by: Edson Alves Filho

When analysing the indicators selected for the watershed system, more precisely for the indicator *Drainage pattern of the watershed, there is* no change when comparing the condition of anthropogenic morphology of active intervention phase with the condition of anthropogenic morphology of consolidated intervention phase: the dendritic pattern seen in the first condition is maintained.

For the indicators *Length of the HiCoom-áfic Basin's Suceficial Perimeter* and *Length of the River Canaia Segments,* the values found for the consensus interment phase are 150 m and 116.40

76

km respectively, which represents an increase of 18 per cent for the first indicator, and 2.23 per cent for the second indicator.

The values found for the indicators *Hydrographic Density of the River Basin, Drainage Density of the River Basin* and *Maintenance Coefficient* showed slight increases in comparison with the active intervention condition, from 6.12% (1.84 rivers/km in the active intervention phase to 2.96 rivers/km in the consolidated intervention phase),84 rivers/km^2 in the active intervention phase to 1.96 rivers/km^2 in the consolidated intervention phase), 7.14% (1.95 km/km^2 in the active intervention phase to 2.10 km/km^2 in the consolidated intervention phase) and 5.16% (476.19 m^2 in the active intervention phase to 502.13 m^2 in the consolidated intervention phase).

In relation to the vegetation *types* and their respective *successional stages,* there was a reduction in the area occupied by the Montane Semideciduous Seasonal Forest from 42.92% of the study area in the anthropogenic morphology situation of the active intervention phase to 36.17% in the anthropogenic morphology phase in the post-intervention phase. The same is true of the Alluvial Semideciduous Seasonal Forest, which occupied 2.98% of the study area in 1998 (Active Intervention) and now occupies 2.96% (Consolidated Intervention Phase). These falls were reflected in the increase in Herbaceous Cover, whose share of the study area in the active intervention phase fell from 40.35% to 44.09%. The same was true of agricultural crops, which increased from 9.34% in 1998 to 10.33% in 2016.

When analysing the indicators relating to climatic conditions, the comparison between the anthropogenic morphology in the active intervention phase and the anthropogenic morphology in the consolidated intervention phase revealed few changes. The values for *average annual temperature, relative humidity, daily precipitation and annual precipitation* showed small positive variations, respectively of 12.19 per cent (down from 18°C during the active intervention phase to 20.5°C during the consolidated intervention phase), 24.40 per cent (down from 50.28 per cent during the active intervention phase to 66.51 per cent during the consolidated intervention phase), 6.44 per cent (down from 13.2 mm per day during the active intervention phase to 14.11 mm during the consolidated intervention phase), -8.68 per cent (down from 1.748 mm per year in the active intervention phase to 1,596.2 mm in the consolidated intervention phase).

As for the indicators relating to the Slope *System, there were* no changes in the area occupied by the slope classes, since the curves used to generate the MDT were the same for the active intervention and consolidated intervention situations. The average height of the slopes in the two situations analysed also remained the same for the same reason (950 m).

For the indicators *orientation* and *average length of the slope ray,* there were no differences between the situation of active intervention and consolidated intervention, due, once again, to the use of the same topographic survey.

With regard to vertebrate *processes*, the formation of furrows, gullies and mass movements were considered. When comparing the active and consolidated intervention phases, there was a slight increase in all the mapped processes, with an increase in values of 3.3 per cent respectively (increase in area from 72.89 ha to 75.42 ha).

In the disturbed *soils* category, the volume and area of mapped cuts and embankments did not increase significantly when comparing the active intervention and consolidated intervention situations. The area and volume of mapped cuts increased by 1.20 per cent (from 224,176.12 m2 in the active intervention phase to 226,905.15 m2 in the consolidated intervention phase) and 4.28 per cent (from 76,177,418.13 m3 in the active intervention phase to 79,586.309.38 m3 in the consolidated intervention phase), as well as the area and volume of landfills, which increased by 0.63% (from 18.90 ha in the active intervention phase to 19.02 ha in the consolidated intervention phase) and 1.70% (from 57,816,414.12 m3 in the active intervention phase to 58,818,241.65 m3 in the consolidated intervention phase) respectively, demonstrating the stabilisation of the interventions of the developments under study in the analysed landscape. The areas of exposed surfaces also increased when comparing the two situations, with a 2.53 per cent increase (from 82.16 ha in the active intervention phase to 84.30 ha in the consolidated intervention phase).

With regard to -*eological -natures, there are* no changes when comparing the active and consolidated intervention situations, since the same survey was used in both situations.

When analysing *erosive* processes, the processes relating to the installation of furrows, gullies and mass movements were mapped, which showed a slight increase when comparing the situations of anthropogenic morphology in the active intervention stage and anthropogenic morphology in the consolidated intervention stage, i.e. 3% (from 72.89 ha during the active intervention phase to 75.16 ha in the consolidated intervention phase). The volumes eroded increased by 0.11% between the two situations analysed (from 252,828,482.70 m3 in the active intervention phase to 253,124,832.17 m3 in the consolidated intervention phase). The frequency of erosion processes mapped was 1.33 processes per km2.

Consultation of the Environmental Impact Study for the Rio do Peixe II SHPP indicated a reforestation area for the reservoir's PPAs of 50 ha, with an estimated revegetation time of 6 years.

For the indicators of the *river system,* more specifically those of the *plains* and terraces *subsystem*, the areas and *perimeters of the river plains* and *river terraces* remained practically the same, with variations of 0.34% (from 204.41 ha to 205.12 ha) and 1.20% (from 51.50 km to 52.13 km) respectively. The spatial distribution of the river plains and terraces didn't change much either, with small increases of 0.94% (from 102.15 ha to 103.12 ha) and 5.11% (from 19.11 km to 20.14 km).

Finally, among the indicators selected for the plains and terraces subsystem, mention should be made of the maintenance of the area and width of *the bodies of water, as* well as the *maximum elevation,* maximum flood area and area of the flood plain suppressed by the reservoir, when comparing the consolidated intervention situation with the active intervention.

For the **canal/reservoir** subsystem, when analysing the water *quality* indicators, it should be noted that it was possible to make a comparison between the 36 indicators listed in the active intervention phase due to the fact that the Rio do Peixe II SHPP Operating Licence Renewal Report was accessed, in which monitoring the water quality of the reservoir and the main watercourses is an essential condition for renewing the licence. When analysing these indicators together and comparing them with the situation of anthropogenic morphology in the active intervention phase, there is an increase in the indicators of total coliforms (58.82% increase), faecal coliforms (93.3% increase), total phosphorus (64.70 increase) and nitrate nitrogen (57.3% increase).

When analysing the set of *gmometrOn ftminl* indicators, the values found in the active intervention situation were maintained for the consolidated intervention condition (Indicators *Main Channel Section, Surface Width, Average Section Depth, Total Reservoir Volume* and Reservoir *Perimeter*). The same situation was found for the indicators *Useful Reservoir Volume, Longitudinal Channel Profile, River Gradient,* Cross-Sectional Area, Wetted Area, Wetted *Perimeter* and *Hydraulic Radius.* Finally, the values found for the Channel Shape Index and the Original Bed Sinuosity Index were maintained.

For the flow indicator, the *main channel flow* and *peak flow* values showed the following behaviour: an increase of 36.80% for Flow Q7.10 in the comparison between the active and consolidated intervention phases; an increase of 37.02% in the average flow in the comparison between the active and consolidated intervention phases and, finally, maintenance of the peak flow values at 14.61 m3/s.

The variations in *the altitude level in the reservoir as a* result of the reservoir's operating regime and the *daily, monthly* and *annual depletion levels* were within the historical average of the monitoring series, which did not lead to significant changes when comparing the active and consolidated intervention phases.

For the *inunPaCtions* indicator, the *recording points, the number of events, the high metric level reached, the duration of events* and the znoxious *flow of events* remained at the historical averages, and therefore at the same values as those recorded for the situation of anthropogenic morphology in the active intervention phase. For design reasons, the *reduced flow* stretch remained the same in the two conditions analysed.

The same conditions were maintained for the *shape* indicators *and the water cure* indicators as in the active intervention condition.

Due to the use of sub-orbital and orbital images to measure geo-indicators in a situation of consolidated intervention (aerial photographs and satellite images), the possibility of inaccuracies in the measurements taken should be emphasised, due not only to the nature of the remote sensing products mentioned, but also to the scales used in the maps that served as the basis for extracting the results found. Complementation by fieldwork is therefore vitally important for checking and correcting possible mistakes, which will make the surveys even more accurate.

Finally, in order to specialise the geo-indicators in the condition of Anthropogenic Morphology in a Consolidated Intervention Phase, **Map 2.4.a** - Anthropogenic Morphology in a Consolidated Intervention Phase in the Study Area was drawn up.

Map 2.4.a was drawn up using Google Earth images with a spatial resolution of 50 cm. Based on Table 2.4.a shown above, only the geo-indicators with spatial manifestation were specialised, in a procedure that repeated the operations carried out for the anthropogenic morphology geo-indicators in the active intervention phase.

The spatial distribution of the categories with punctual, linear and areal manifestations on Map 2.4.a was practically the same as that seen on Map 2.3.f. The only exceptions were the number of tracks referring to cultivation areas, which were more numerous in the consolidated intervention phase, due to the increase in the cultivation area in the latter condition, as well as the area represented by the distribution of macro-tracks in the Rio do Peixe II SHPP reservoir, which was more effective in the consolidated intervention condition.**Map 2.4.a** - Anthropogenic Morphology in the Consolidated Intervention Phase is presented below:

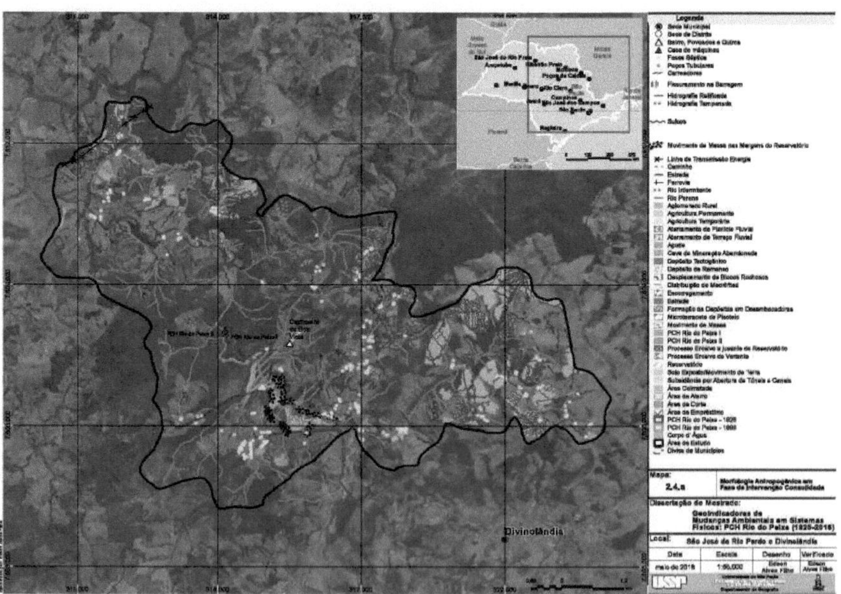

Map 2.4.a - Anthropogenic Morphology in Phase of Consolidated Intervention in the Study Area

The following are some comments that take into account the comparison between the Integrated

Environmental Assessment of the Physical Environment methodology for hydroelectric projects and the methodology proposed by the Anthropological-Morphological Approach, which uses Detailed Geomorphological Mapping as its main input.

2.5 Comparison between the Survey and Mapping Approaches

Negative Environmental Impacts on the Physical Environment by the Electricity Sector with the possibilities opened up by the Anthropogeomorphological Approach

As presented in **Section 2.1** of this volume, the Integrated Environmental Assessment methodology has become an important tool for assessing the environmental viability of hydroelectric projects in Brazil, and has been used by the Eletrobrás System as a tool for selecting alternatives for splitting the falls in various Brazilian regions.

In this methodology, the assessment of environmental sensitivity would be responsible for indicating the natural limitations that environmental systems show in their own dynamics. In the assessment of socio-environmental impacts, this methodology selects the impacts based on the impacting actions derived from the different phases of work common to hydroelectric projects, and from there assesses the incidence, scope, probability, significance, importance, cumulativeness and synergy, resulting from the sum and multiplication of factors between these variables the final value of the impact and its normalisation in values from 1 to 5.

Based on the assessment and measurement of socio-environmental impacts, it would be possible, within the integrated environmental assessment methodology, for a new configuration of environmental attributes to emerge after the implementation of a given hydroelectric project, measured by means of the arithmetic composition between the values found for environmental sensitivity and environmental impacts, generating what the methodology defines as environmental fragility.

When analysing the cartography generated from Map 2.1.m, there is a tendency to homogenise the regions indicated as having the greatest cumulative impacts. The cartographic solution of supporting the manifestation of impacts by areas of coverage, calculating their values by sum and product, ends up masking small variations in magnitude between the variables found. An example of this is the relative impact of surface water quality, which, in the mapping carried out in Map 2.1.m, is represented by the colour orange. For all we know, it shows variations over time and over the river's flooding cycles, which is not possible to capture with the proposed methodology.

Although the sensitivity mapping is intended to demonstrate the peculiarities of the physical systems throughout a river basin, showing the areas with the greatest restrictions when considering the systemic particularities of each variable, when comparing this mapping with the detailed morphological mapping developed for the situation of semi-preserved original morphology (Map 2.2.a), the use of segmentation of the compartments of the physical environment, such as that used by this detailed geomorphological mapping technique, makes it possible to read the variation in morphometric and morphographic attributes throughout the landscape, giving a good idea of how the physical systems in the study area are organised.

Although this mapping does not show a scale of value in its legend, the quantifications generated from it allow for a qualitative analysis of the organisation of the landscape in the study area before the implementation of the Rio do Peixe I SHPP.

The methodology for mapping environmental sensitivities, as recommended by the Integrated Environmental Assessment procedures, does not make it possible to understand and delineate short-cycle environmental changes, since it qualifies, in degrees of sensitivity, groups of the physical environment whose origins and processes are of the most diverse ages, thus homogenising all the compartments analysed.

Generally, when a hydroelectric project is installed in a given location, it is precisely the geomorphological parameters and systems whose processes are triggered in short time-scale cycles that are the first to be altered by the intervention. Far from trying to detract from the great merit of the methodology for mapping environmental sensitivities, what we wanted to say is that mapping the original morphology can be added to it as a next step, providing a better diagnosis of the situation prior to the implementation of the project, and even allowing new variables to be added to the list of impacts. This provides a very well-defined framework for understanding the magnitude of the changes in environmental alterations.

When the Integrated Environmental Assessment methodology makes its final feasibility assessment for the installation of hydroelectric projects, it combines the sensitivity and impact maps by means of averages and weights to generate the final prognosis by means of fragility. In this process, it ends up giving greater weight to the areas located close to the reservoir, which only accentuates local impacts, giving less emphasis to the other physical compartments, such as the slopes and dividers.

By selecting geo-indicators for all geomorphological compartments from the pre-intervention phase, it is possible to make comparisons for a whole area, understanding the magnitude of environmental changes in all physical compartments and in all planning phases of a given intervention.

CHAPTER 3

FINAL CONSIDERATIONS

The application of part of the ideal list of indicators and geo-indicators for the study area made it possible to quantify and qualify, as well as understand, the extent of the environmental changes resulting from the implementation of the Rio do Peixe I and II SHPs.

With regard to indicators, they were restricted to topics such as water quality, flood levels, physical aspects of river bed materials, chemical and physical aspects of soils and rocks, water quality and flow and flood data, broken down into all the physical systems and subsystems analysed, mentioning, of course, the existence of information gaps due to the fact that it was not possible to find data for all the parameters. The geo-indicators include parameters such as morphometric data from the analysis of river basins, vegetation cover, rock types, soils, deforestation data, erosion, APP revegetation, morphological data from plains and river terraces, among others, also with data gaps due to the fact that no technical studies could be found to extract all the information listed in the geo-indicator list.

The mapping of land use and associated morphological interferences quantified and demonstrated the direction of alteration of the landscape studied, - *the* geoindicators relating to river geometry, water quality, materials and processes of the slope and river plain, obtained in the studies that subsidised the environmental licensing and renewal of licences for the projects. These elements made it possible to gain a direct understanding of the extent of the alterations made to the landscape, in a universe of analyses spanning more than 90 years.

When only the geo-indicators for which there is data in the three intervention phases analysed are analysed, as shown in **Table 3.0.a below, we** initially went from a complete list of geo-indicators with 211 indicators to a list with 143, for which the documents consulted and mapping carried out during the research produced data. However, by systematising the data found for those that actually manifested themselves in the three morphological conditions studied, the value of the indicators drops to 37, which effectively allow us to know the magnitude of the landscape derivation process in the study area, as they manifest themselves in the pre-intervention conditions (semi-preserved original morphology) and in the active and consolidated intervention conditions (anthropogenic morphology).

Table 3.0.a - Assessment of the Magnitude of Environmental Impacts in the Study Area through the Evaluation of Geoindicators in the Condition of Semi-Preserved Original Morphology in the Pre-Intervention Phase and in the Condition of Anthropogenic Morphology in the Active and Consolidated Intervention Phases (1930-2016)

			Results		
System	**Indicator**	**Parameters**	Semi-preserved original morphology	Anthropogenic Morphology	

			Pre-Intervention Phase	Active Intervention Phase	Consolidated Intervention Phase
Basin Hydrographic	Morphological/ Morphometric	Drainage pattern of the catchment area	Dendritic to Subdendritic	Dendritic	Dendritic
Basin Hydrographic	Morphological/ Morphometric	Length of the river basin's surface path$^{(Eps)}$	150 m	23 m	23 m
Basin Hydrographic	Morphological/ Morphometric	Length of Channel Segments (C.s.)	171.56 kilometres	113.80 kilometres	113.80 kilometres
Basin Hydrographic	Morphological/ Morphometric	Drainage Area	53.89 kilometres^2	53.89 kilometres^2	53.89 kilometres^2
Basin Hydrographic	Morphological/ Morphometric	Hydrographic density of the catchment area (Dh)	4.60 rivers/km^2	1.84 rivers/km^2	1.84 rivers/km^2
Basin Hydrographic	Morphological/ Morphometric	Drainage Density of the Basin Hydrographic (Dd)	3.13 km/km^2	2.10 km/km^2	2.10 km/km^2
Basin Hydrographic	Morphological/ Morphometric	River Basin Maintenance Coefficient (Cm)	310 m /m^2	476,19 m^2	476,19 m^2
Basin Hydrographic	Materials	Type of Cover Vegetable	Montane Semideciduous Seasonal Forest (27.70 km^2), Alluvial Semideciduous Seasonal Forest (1.85 km)^2	Montane Semideciduous Seasonal Forest in an Advanced Stage of Regeneration (0.35 km^2), Seasonal Forest Semideciduous Montane Forest in Medium Stage of Regeneration (8.42 km^2), Seasonal Forest Semideciduous Montane Forest in the Initial Stage of Regeneration (12.08 km^2), Seasonal Forest Semideciduous Montane Forest in the Pioneer Stage of Regeneration (2.35	Montane Semideciduous Seasonal Forest in Stage Advanced Regeneration Stage (0.57%), Montane Semideciduous Seasonal Forest in Medium Regeneration Stage (13.37%), Montane Semideciduous Seasonal Forest in Initial Regeneration Stage (18.22%), Montane Seasonal Forest in Pioneer Regeneration Stage (4.01%), Alluvial Semideciduous Seasonal Forest in Advanced Stage of Regeneration (0.0001%), Alluvial Semideciduous Seasonal Forest in Medium Stage of Regeneration (0.88%), Alluvial Semideciduous

				km²), Seasonal Forest Semideciduous Alluvial in Medium Stage of Regeneration (0.45 km²), Seasonal Semideciduous Forest Alluvial Forest in the Initial Stage of Regeneration (1.01 km²), Alluvial Semideciduous Seasonal Forest in the Initial Stage of Regeneration (1.01 km) Pioneiro de Regeneração (0.14 km).²	Seasonal Forest in Initial Stage of Regeneration (1.81%), Alluvial Seasonal Forest in Pioneer Regeneration Stage (0.27%)
Basin Hydrographic	Materials	Successional Stage of Cover Vegetable	Analysing aerial photographs from 1930, the study area was considered to be covered by Seasonal Forest Semideciduous Montane Forest in an Advanced Stage of Regeneration, as well as Seasonal Forest Semideciduous Alluvial Forest in an Advanced Stage of Regeneration	Pioneer Regeneration Stage (2.49 km²), Initial Regeneration Stage (13.09 km²), Medium Regeneration Stage (8.87 km²), Stage Advanced Regeneration (0.35 km).²	Advanced Stage of Regeneration (31.19 ha), Middle Stage of Regeneration (61.65 ha), Initial Stage of Regeneration (1,096.06 ha), Pioneer Stage of Regeneration (234.13 ha)
Basin Hydrographic	Process	Daily Precipitation	8.91 mm/day (January - wettest month)	410 mm (December, wettest month)	437.5 mm (January, wettest month)
Basin Hydrographic	Process	Annual Precipitation	1,201.9 mm	1,748 mm	1,596.2 mm
System Strand	Process	Area of Occupation of Slope Classes	4.97 km² (up to 2%), 5.23 km² (2% to 8%), 10.82 km² (8% to 15%), 23.88 km² (15% to 30%), 8.66 km²	4.97 km² (up to 2%), 5.23 km² (2% to 8%), 10.82 km² (8% to 15%), 23.88 km² (15% to 30%), 8.66 km² (30% to 45%), 1.14 km² (>	4.97 km² (up to 2%), 5.23 km² (2% to 8%), 10.82 km² (8% to 15%), 23.88 km² (15% to 30%), 8.66 km² (30% to 45%), 1.14 km² (> 45%)

			(30% to 45%), 1.14 km^2 (> 45%)	45%)	
System Strand	Process	Altitude Strand average	950 m	950 m	'950 m
System Strand	Process	Guidance Predominant	4.96 km^2 (flat), 6.16 km^2 (north), 4.22 km^2 (north-east), 5.50 km^2 (east), 6.02 km^2 (south-east), 5.39 km^2 (south), 4.46 km^2 (south-west), 8.52 km^2 (west), 9.47 km^2 (north-west),	4.96 km^2 (flat), 6.16 km^2 (north), 4.22 kilometres^2 (north-east), 5.50 km^2 (east), 6.02 km^2 (south-east), 5.39 km^2 (south), 4.46 km^2 (south-west), 8.52 km^2 (west), 9.47 km^2 (north-west),	4.96 km^2 (flat), 6.16 km^2 (north), 4.22 km^2 (north-east), 5.50 km^2 (east), 6.02 km^2 (south-east), 5.39 km^2 (south), 4.46 km^2 (south-west), 8.52 km^2 (west), 9.47 km^2 (north-west),
System Strand	Process	Average Ramp Length	2.100 m	2.100 m	2.100 m
System Strand	Materials	Units Geological	10.95 km^2 (migmatites) of the Varginha-Guaxupé Group), 16.12 km^2 (paragneisses of the Varginha-Guaxupé Group), 27.61 km^2 (charnockites of the São José do Rio Pardo-Divinolândia Group).	10.95 km^2 (migmatites of the Varginha- Group) Guaxupé), 16.12 km^2 (paragneisses of the Varginha-Guaxupé Group), 27.61 km^2	10.95 km^2 (migmatites of the Varginha-Guaxupé Group), 16.12 km^2 (paragneisses of the Varginha-Guaxupé Group), 27.61 km^2 (charnockites of the São José do Rio Pardo-Divinolândia Group).
River/Lake System	Morphological/ Morphometric	Total Area of Water Bodies	20 ha	125.44 ha	125.44 ha
River/Lake System	Morphological/ Morphometric	Total Perimeter of Water Bodies	14.12 kilometres	60.65 kilometres	60.65 kilometres
River/Lake System	Morphological/ Morphometric	River Plain Area	262 ha	204.41 ha	204.41 ha
River/Lake System	Morphological/ Morphometric	Perimeter of the River Plain	70.48 kilometres	51.50 kilometres	51.50 km
River/Lake System	Morphological/ Morphometric	Distribution of the River Plain	It is distributed in the study area in the S	It is distributed in the study area in the S-NO direction	It is distributed in the study area in the direction S-NO

			direction -NO		
River/Lake System	Morphological/ Morphometric	River Terrace Area	93 ha	102.15 ha	102.15 ha
River/Lake System	Morphological/ Morphometric	Perimeter of River Terraces	16.71 kilometres	19.11 kilometres	19.11 kilometres
River/Lake System	Morphological/ Morphometric	Distribution of River Terraces	It is distributed in two bands in the study area: one to the NE, downstream of the Rio do Peixe I and II SHPs, and one to the S, just upstream of the Rio do Peixe I and II SHPs.	It is distributed in three areas: the first to the NW, near the confluence of the Peixe River with the Pardo River, the second, to the N, located 1.4 km from the first area, and finally, the third, to the SO, starting from the reservoir of the Peixe River SHPs. I and II.	It is divided into three areas: the first to the NW, near the confluence of the River Peixe with the River Pardo, the second to the N, located 1.4 kilometres from the first area, and finally the third, to the SO, starting from the reservoir of the Rio do Peixe I and II SHPs.
River/Lake System	Process	Main Channel Flow (Q7, Q10)	8.34 m^3/s	11.75 m^3/s (average) Q7.10 (1.03 m^3/s), Long Term Average Flow (7.26 m^3/s)	Long-term average flow (7.4 m^3/s), Q7.10 flow (1.63 m^3/s)
River/Lake System	Process	Peak Flow	73.50 m^3/s	14.61 m^3/s	14.61 m^3/s
River/Lake System	Morphological/ Morphometric	Width Channel Section (Wmp)	22,50 m	22,50 m	22,50 m
River/Lake System	Morphological/ Morphometric	Width Surface (L)	12,25 m	12,25	12,25 m
River/Lake System	Morphological/ Morphometric	Average Section Depth (Dmp)	2,20 m	2,20 m	2,20 m
River/Lake System	Morphological/ Morphometric	Profile Longitudinal Channel	Characteristic equilibrium profile with knick points at the confluence with higher order channels.	Characteristic equilibrium profile with knick points at the confluence of higher order channels.	Characteristic equilibrium profile with knick *points* at the confluence of higher order channels.
River/Lake System	Morphological/ Morphometric	River Gradient	24,52 m	24,52 m	24,52 m
River/Lake System	Morphological/ Morphometric	Cross-sectional area (Amp)	19,31 m^2	19,31 m^2	19,31 m^2
River/Lake System	Morphological/ Morphometric	Wet Area (A)	19,28 m^2	19,28 m^2	19,28 m^2
River/Lake System	Morphological/	Perimeter Wet (P)	26,90 m	26,90 m	26,90 m

	Morphometric				
River/Lake System	Morphological/ Morphometric	Hydraulic Radius (R)	0.71 m/m^2	0.71 m/m^2	0.71 m/m^2
River/Lake System	Morphological/ Morphometric	Channel Shape Index (F)	0,26	0,25	0,25
River/Lake System	Morphological/ Morphometric	Sinuosity Index (Is)	2,50	2,50	2,50

Prepared by: Edson Alves Filho

The above compilation, which shows the figures for the geo-indicators surveyed for the pre-intervention, active intervention and consolidated intervention phases, revealed a sharp decrease in the Watershed Surface Path Length (SWTL) when comparing the pre-intervention phase with the active intervention and intervention phases.

consolidated. The reduction observed was 84.6 per cent, from 150 m during the pre-intervention phase to 23 m during the active intervention and post-intervention phases.

The Surface Runoff Length is an index that shows *how* far surface runoff flows until it reaches a watercourse, contributing to an increase in the volume of water in the receiving watercourses. The decrease in values *is* due to the reduction in the area available for water infiltration into the soil, which is directly related to the implementation of the Rio do Peixe II SHPP reservoir and the increase in agricultural areas in the study area.

Another indicator to show major changes in the morphological conditions analysed is the Length of River Channel Segments, with a 33.66% reduction between the pre-intervention situation and the active and consolidated intervention situations, when it went from 171.56 km to 113.80 km. This decrease is related to the process of drying up the river channels, due to the process of installing the reservoir in the study area, the consequences of which on subsurface and deep drainage is to alter the groundwater recharge cycle, reducing the water available for recharging the watercourses, which causes the position of springs and perennial watercourses to change over time.

As a result of the installation of the Rio do Peixe II SHPP reservoir, the Hydrographic Density of the Hydrographic Basin decreased, due to the drying up of the river channels as a result of the alteration of the underground water flow. In the pre-intervention phase, the hydrographic density of the river basin was 4.60 rivers/km2, a figure that fell by 60 per cent in relation to the active and consolidated intervention phases, when it reached 1.84 rivers/km^2 .

With the decrease in the number of perennial watercourses due to the implementation of the Rio do Peixe II SHPP reservoir, another morphometric index to suffer a strong change in the morphology conditions analysed is Drainage Density, with a 32.90% decrease between the pre-intervention phase and the active and consolidated intervention phases, when it drops from 3.13 km/km^2 to 2.10 km/km^2

As a result of the decrease in the basin's infiltration surface due to the installation of the Rio do Peixe II SHPP reservoir, the Maintenance Coefficient, which is the minimum drainage area needed to maintain 1 m of surface runoff in the watercourses, increased dramatically. There was an increase of 34.91%, from an area of 310 m^2 to maintain 1 m of surface runoff in the pre-intervention phase, to an area of 476.19 m^2 to maintain 1 m of surface runoff in the basin's watercourses.

The total area of water bodies within the fluvial/lacustrine system has increased by 84.05%, caused exclusively by the implementation of the Rio do Peixe II SHPP reservoir. As a result of the increase in the length of the water bodies in this relief compartment, there was an increase in the perimeter of the water bodies of around 76.71%.

Another reflection of the implementation of the Rio do Peixe II SHPP is the reduction in the area and perimeter of the river plain in the study area, with reductions of 21.98% and 26.92% respectively when comparing the pre-intervention phase with the active and consolidated intervention phases.

The downstream erosion effect caused by the installation of the Rio do Peixe II SHPP reservoir may be one of the reasons for the increase in the area of river terraces in the study area. Due to the installation of the reservoir, the dam retains the load of suspended sediment coming from upstream in the basin, accumulating it at the bottom of the reservoir. As a result, the water coming out of the plant's spillway is poor in sediment, which makes its erosive potential very high, causing it to erode the stretches of river plain located downstream of the dam, making the water again rich in suspended sediment, which is deposited in stretches of plain further downstream, contributing to the formation of new levels of river terrace. The increase in the area and perimeter of the river terraces between the pre-intervention phase and the active and consolidated intervention phases was 8.9% and 12.55% respectively.

Finally, with regard to the flow of the main channel, represented by the course of the River Peixe in the study area, there was a decrease of around 87.64% due to the installation of the Rio do Peixe II SHPP reservoir (a decrease between the pre-intervention phase and the active intervention phase). After the reservoir was installed, there was an increase of 36.8% (between the active intervention phase and the consolidated intervention phase) due to the operational rules for maintaining the reservoir level, as evidenced by the operational reports contained in the Rio do Peixe II SHPP Operating Licence Renewal Report. The same behaviour can be seen with Peak Flows, which fell by around 80.12% between the pre-intervention phase and the active and consolidated intervention phases, due to the flood regulating effect caused by the installation of reservoirs.

Generally speaking, when we look at the behaviour of the geo-indicators for the three phases considered (pre-intervention, active intervention and consolidated intervention), we see a break in the values between the pre-intervention and active intervention phases, with large variations in the values considered, and a maintenance in the values between the active intervention and consolidated intervention phases, revealing that the agent responsible for the greatest transformations in the landscape studied is hydroelectric use. One explanation for this behaviour lies in the fact that many of the geo-indicators used were taken from technical reports, such as the Rio do Peixe II SHPP Environmental Impact Report and the Rio do Peixe II SHPP Operating Licence Renewal Report, which only presented data for 1998 and 2007, It was not possible to find studies with values for other years within the intermediate period, which would make it possible to see whether the situation of accommodation of the different elements that make up the landscape in the study area behaved in this way in the period between the active intervention phase and the consolidated intervention phase.

In general, the genetic significance of the disruption of landscape processes reported by the geo-indicators that gather data for the three phases of analysis considered (pre-intervention, active intervention and post-intervention) allow us to verify that the installation of the Rio do Peixe II SHPP represented a disruption in the surface hydrological process of the area analysed, causing changes in sub-surface and underground hydrological functioning that had repercussions on surface hydrological behaviour. The drying up of river channels due to the buffering effect of the infiltration surface caused by the installation of the Rio do Peixe II SHPP has displaced the feeding points of the river channels, with effects on drainage density and the maintenance coefficient. Alterations to these parameters cause changes to the geometry and balance of the slopes, changing the balance of erosion processes within the basin analysed, which has a long-term effect on the stability of the forms and materials of the relief analysed. Another effect is a reduction in the capacity to drain rainfall discharges of great intensity and short duration, which can increase the occurrence of mass movements and floods.

It should be noted, however, that the agricultural history of the study area, which makes it possible to highlight agricultural use as far back as the 20th century, already appeared in the form of patches on the Physical Land Use Categories Map in 1930, and increased over the 91-year time series analysed.

As NIR (1983) points out, agricultural use has notorious effects on the balance of pedogenesis in a given area, increasing soil loss due to incorrect management practices. In the study area, agriculture's role in morphogenesis is quite significant, both in terms of contributing to the generation of sediment that will be deposited in the Rio do Peixe II SHPP reservoir, and in contributing to the drying up of river channels due to the existence of impoundments, such as dams, which help to alter the underground hydrological behaviour of the entire area.

However, as shown by the indicators and geo-indicators worked on, as well as the land use figures taken from the Maps of Physical Land Use Categories, agricultural use, although old in the study area, was residual in 1930, since the landscape matrix was mostly forested.

Agricultural uses spread across the study area gradually over the course of 91 years, while the first hydroelectric intervention, dating back to 1925, brought a concentrated force of intervention in just a few years. The same happened in 1998, when the Rio do Peixe II SHPP was implemented, when another concentrated moment of intervention on the physical environment took place in the study area.

Although, in this second case, the cumulative history of agricultural interventions was already significant, without the action of hydroelectric use in damming the river plain and changing the hydrological dynamics in the study area more incisively, it would not have been possible to potentialise the very landscape changes caused by agricultural use.

Mass movements, dams, cuts, embankments and furrows would not have been formed in such a widespread way in the study area, and at an even greater rate, if an element of great morphological disturbance to the physical environment, such as hydroelectric use, had not been implemented.

The choice of indicators and geo-indicators for the study area included those relating to the plain, river terrace, lake and canal system, which more clearly reflect the impact of hydroelectric use in the study area. As for the indicators related to the slope system, the impact of agricultural use is more noticeable, since cuts, embankments, dams and mass movements were caused by the installation of farms and ranches.

However, it would be reasonable to expect that the anthropogenic changes in the landscape caused by agriculture in the study area would not have such intense effects on sedimentation and surface dynamics in the relief without the installation of a larger disturbance whose effects are felt on a regional scale, such as hydroelectric use.

Bearing in mind the implications of the changes produced by the implementation of hydroelectric use in the study area, it can be seen that - *a* reduction in the maintenance coefficient will have an effect on the operational safety of the Rio do Peixe II SHPP, since the need for larger areas to capture precipitation in the study area will have an effect on the production of water to flow into the reservoir, with a consequent reduction in the water available for the flow of the Rio do Peixe, and greater dependence on flows from upstream to the dam to fill the reservoir and maintain the reservoir required to maintain the plant's operational safety.

The current situation of climatic oscillations that have had an effect on altering rainfall in the basin over the period analysed, such as the accentuation of drought episodes, as occurred in 2015, will require measures for better monitoring of flow conditions in the study area, as well as the reinforcement of environmental compensation measures, since the environmental compensation applied in the study area resulted in the planting of eucalyptus in some areas near the reservoir.

Perhaps encouraging the legalisation of the legal reserves of the numerous rural properties in the study area, which use almost all of their land for agricultural crops, would increase the area of vegetation cover in the study area, increasing water production and mitigating the hydrological imbalances caused by the implementation of the Rio do Peixe II SHPP.

Another action could be the implementation of an Environmental Conservation and Use Plan for the Surroundings of the Artificial Reservoir (PACUERA), a new environmental compensation mechanism for the Brazilian electricity sector, which has not been implemented in relatively old plants such as the Rio do Peixe II SHPP in 1998. It proposes zoning for the reservoir's surroundings, where vegetation cover is recovered with native species to avoid the effects of erosion along the reservoir.

The environmental compensation applied to the Rio do Peixe II SHPP only implemented a few areas of eucalyptus plantations in the reservoir area, an exotic species that is not very effective at protecting the reservoir's banks, and most of the reservoir's perimeter is currently covered in herbaceous vegetation and leisure farm areas.

The increase in river plains and terraces over the historical period analysed also demonstrates the effectiveness of the disruption of natural processes caused by hydroelectric use in the study area. The introduction of two dams over 91 years changed the direction of sedimentation in the study area, with an increase in the level of the terraces downstream of the dam, near the confluence of the Peixe River with the Pardo River.

In this way, the landscape of the study area has gained sandy bars and even the encroachment of former regions that were dedicated to horticulture in older mappings (1930, 1971), verifying - progressive occupation by herbaceous vegetation. Throughout Quaternary environmental history, where the study region gradually moved from a dry climate to a typical tropical climate, albeit marked by seasonality, rain erosion was responsible for generating rounded shapes and watercourses in an equilibrium profile. Over the 91 years analysed, however, according to *the* geoindicator analysis carried out, there have been changes in the profiles of the slopes, caused by both hydroelectric and agricultural use, which places the anthropic factor as an element in the derivation of natural processes within the environmental history of the study area.

When analysing the results generated for this research, especially when comparing the results of the mapping and survey of geo-indicators based on the theoretical-methodological references of Detailed Geomorphological Cartography and Anthropogeomorphology with the mapping resulting from the methodological guidelines of the Integrated Environmental Assessment of Hydroelectric Projects, carried out by the Eletrobrás System, one can see a wealth of parameters and understanding of the environmental change of the attributes of the physical environment in the former to the detriment of the latter.

The tendency towards homogenisation of weights and averages and the selection of impact coverage areas brought about by the theoretical-methodological framework of the Integrated Environmental Assessment of Hydroelectric Projects leads to a greater valuation of local negative environmental impacts to the detriment of regional negative environmental impacts.

By adopting the methodological option created by the Anthropogeomorphological Approach, in which there is a direct selection of geo-indicators by means of maps drawn up for the conditions of semi-preserved and anthropogenic original morphology, the impacts are mapped directly by the physical evidence of their manifestations. Their occurrence is also differentiated in relation to the stage of implementation of the hydroelectric project studied. This produces, firstly, a more sophisticated and precise cartographic result, and secondly, a more accurate quantification of the magnitude of the impacts considered, than that seen by simply multiplying variables, as shown by the methodology used by the Brazilian electricity system.

CHAPTER 4

BIBLIOGRAPHICAL REFERENCES

AB'SABER, A. N. **A Terra Paulista**. *In:* Boletim Paulista de Geografia, São Paulo, n. 23, p. 5-38, 1956.

AB'SABER, A. N. **Potencialidades Paisagísticas Brasileiras**. *In:* Boletim Geomor- fologia, São Paulo, Instituto de Geografia da USP, n. 55, 1977.

ABREU, S.F. **Recursos Minerais do Brasil,** 2.ed, São Paulo, Edgard Blucher/ Rio de Janeiro, Instituto Nacional de Tecnologia, 1973.

ALMEIDA, F. F. M. de. **Geological Foundations of the Paulista Relief.** São Paulo: USP Geography Institute, 1964. 99 p. Series of Theses and Monographs, 14.

ALVAREZ, J. A. M. 1986. **Reservoir siltation and bed erosion downstream of dams.** *In:* Latin American Hydraulics Congress, XII. São Paulo.

BAILLY, A.; BÉGUIN, **H. Introduction à la Géographie Humaine.** Paris: Masson, 1982.

BARBER, Lynn. **The heyday of Natural History: 1820-1870.** Garden City, New York, Doubleday, 1980.

BERGER, A. R. **The Geoindicator Concept and its application: an introduction.** *In:* BERGER, A. R; IAMS, W. J. (ed). Geoindicators: assessing rapid environmental changes in Earth systems. Rotterdam: A. A. Balkema/Brookfield, 1996.

BERMAN, Marshall. **All that is solid vanishes into thin air: the adventure of modernity.** São Paulo: Companhia das Letras, 1986.

BERTRAND, G. **Paisagem e Geografia Física Global: esboço metodológico.** Caderno de Ciências da Terra, São Paulo, no. 13, 27 p., 1971.

BIGARELLA J. J. & BECKER R. D. 1975. **Sea level changes**. In: International Sym- posium on the Quaternary. Curitiba. Topics for discussion. Boletim Paranaense de Geociências n. 33, pp. 245-251.

BOULDING, K. E. Ecodynamics: A New Theory of Societal Evolution. Beverly Hills, California: Sage. 1978.

BOUNEAU, Christophe & VARASCHIN, Denis. Les paysages de l'électricité: Pers- pectives historiques et enjeux contemporains (XIXe-XXIe siècles). Brussels, Bern, Berlin, Frankfurt am Main, New York, Oxford, Wien, 2012. 273 p., 43.

BRIERLEY, G.; FRYIRS, K. A.; JAIN, V. **Landscape connectivity: the geographic basis of geomorphic applications**. Area, v. 38 (2), p. 65-174, 2006.

BRUNSDEN, D. Geomorphological events and landforms change. In: Zeichtirift fur geomorphology, No. 40, pp. 273-288, 1996.

BRUNSDEN, D.; THORNES, J. B. Landscape Sensitivity and Change. Transactions of the Institute of British Geographers, New Series, v. 4, n. 4, p. 463-484, 1979.

CAILLEUX, A. La ligne de cailloutis a la base des sols jaune. Zeitsch. f.. Geomorph., Band 1, 312 (transcribed Not. Geomorf. 4, p.43, 1957).

CASSIRER, E. **Las ciencias de la cultura**. Mexico: Fondo de Cultura Econômica, 1982.

CATTON JR., W. R. Overshoot: the ecological basis of revolutionary change. Urbana, Illinois: University of Illinois Press, 1980.

CEMA ENVIRONMENTAL CONSULTANCY. Environmental Impact Report (EIA-RIMA) for the Rio do Peixe Power Station. São José do Rio do Pardo and Divinolândia, São Paulo, 1993.

CHORLEY, R. J. **Geomorphology and General Systems Theory. US. Geological** Survey Professional Paper, vol. 500-B, 1962, pp. 1-10.

CHRISTOFOLETTI, A. **Systems Analysis in Geography**. São Paulo: Hucitec, 1979.

CHRISTOFOLETTI, A. **Geomorphology**. São Paulo: Edgard Blucher, 1980.

CLARO, Mariana Sgarbi. **Complex Morphological Units in the Tapera Stream Watershed, São Paulo - SP**: contributions to urban and environmental planning. Master's dissertation. São Paulo: FFLCH/USP, 2013.

COLTRINARI, L. **Physical Geography and Environmental Change.** In: CARLOS, A. F. A (org.). Novos Caminhos da Geografia. São Paulo: Ed. Contexto, Iª . ed., 1999.

COLTRINARI, L. **Detailed Geomorphological Cartography: the graphic representation of relief between 1950 and 1970.** In: Revista Brasileira de Geomorfologia, 2011.

COLTRINARI, L.; MCCALL, G. J. H. **Geoindicators: earth sciences and environmental change.** In: Revista do Departamento de Geografia - USP, São Paulo, no.9, pp. 5-12, 1995.

CONTI, José Bueno. **Secondary Circulation and Orographic Effect in the Rainfall Genesis of the Lesnordeste Paulista Region.** Doctoral **Thesis.** Theses and Monographs Series, no. 18, São Paulo: IGEOUSP, 1975.

COOKIE, R. U.; DOORNKAMP, J.C. Geomophology in Environmental Manage- ment: an introduction. Oxford: Clarendon Press, 1974.

CO-OPERATIVE FOR TECHNOLOGICAL AND INDUSTRIAL RESEARCH AND SERVICES (CPTI). **Basin Plan for the Pardo Water Resources Management Unit (UGRHI-4): Technical Report.** São Paulo: CPTI, 2008. 363p.

CORDANIU. G., MILANIE. J., THOMAZ FILHO. A., CAMPOS D. A. (2000). Tec- **tonic Evolution of South America.** International Geological Congress, Rio de Janeiro.

CPRM. GEOLOGICAL SERVICE OF BRAZIL Programme of basic geological surveys of Brazil: Sheet Santos, Guaratinguetá, Ilha Grande, Campinas, scale 1:250000. Geological Integration Project - Geological Survey of Brazil - Superin Regional trend of São Paulo, 1999.

CROZIER, M. J. **Landslides; causes, consequences, and environment.** Croom Helm, London, 252p, 1986.

CRUMLEY, Carole. Historical Ecology: Cultural Knowledge and Changing Lands- capes. School of American Research Press, 1987.

DAILY. H. E. **Steady-state economics**. San Francisco: Freeman, 1977.

DAMIANI, Amélia Luisa *et al.* **Space at the End of the Century: the new rarity.** 1ª . ed. São Paulo: Contexto, 1999. v. 1. 220 p.

DE MARTONNE, E. 1940. **Problemes morphologiques du Bresil tropical atlantique.** In: **Annales de Geographie, 49: 1-27 and 106-129. Trad. Revista Brasileira de Geografia,** 5:523550, 6:155-178,1943/1944.

Demek, J. **Generalisation of geomorphological maps.** In: Progress Made in Geomor- phological Mapping. Brno, 35-66, 1967.

DINIZ, Renato. **100 years of history and energy.** São Paulo: Via das Artes, 1ª ed., 2012.

DORANTI, C. - 2006. Landscape structure in eastern São Paulo and south-western Minas Gerais: Fission Track Thermochronology and Geomorphology. 106f. Master's thesis Paulista State University, Rio Claro-São Paulo.

DRUMMOND, J. A. Environmental history: themes, sources and lines of research. In: Estudos Históricos, v.4, n.8, p.177-97, 1991.

ELETROBRAS. Centrais Elétricas Brasileiras **Manual de Inventário Hidrelétrico de Bacias Hidrográficas.** Brasília: ELETROBRAS, 2007.

EPE. Energy Research Company. **Integrated Environmental Assessment of Hydroelectric Projects in the Rio Branco Basin.** Hydros Engenharia Ltda. São Paulo, 2011.

ERBERT, H. **Occurrence of granulite facies in southern Minas Gerais and adjacent** areas, **depending on the orogenic structure: hypotheses on its origin.** In: Academia Brasileira de Ciências, 40 (supl.), pp. 215-229, 1968.

ERHART, H. **La theorie bio-rexistique et les problèmes biogéographiques et palaeobiologiques.** Soc.Biogeogr. France, CNRS (288):43-53, 1956.

FORNASARI FILHO, N. *et al.* **Alterations to the physical environment resulting from engineering works.** São Paulo, Instituto de Pesquisas Tecnológicas, 1992 (IPT Publication; n. 1.972).

FUJIMOTO, Nina Simone Vilaverde Moura. **Urban Environmental Analysis in the Metropolitan Area of Porto Alegre-RS: Diluvio Stream Sub-Basin.** PhD Thesis - Faculty of Philosophy, Letters and Human Sciences, University of São Paulo, 2001.

FUNDAÇÃO DE AMPARO À PESQUISA DO ESTADO DE SÃO PAULO. **História da Energia Elétrica no Estado de São Paulo - Projeto Eletromemória I (1960 - 2000).** São Paulo, 2014.

FOREST FOUNDATION. **Estimation of Woody Biomass for different phytophysiognomies in the state of São Paulo.** São Paulo, 2008.

GEORGE, Pierre. **Sociology and Geography.** Rio de Janeiro: Companhia Editora Forense, 1969.

GUPTA, A. The changing geomorphology of the humid tropics. In: Geomorphology, No, 07, pp. 165-186, 1993.

HACK, J. T. **Interpretation of Erodite Topography in** humid **temperate regions.**
In: Notícia Geomorfológica, no. 12, p. 3-37, 1972.

HACKETT, Brian. **Landscape Planning.** Oriel Press, New Castle, 1971.

HART, M. G. **Geomorphology pure and applied. London: George Allen & Unwin,** 1986. 228p.

HESSEN, Johannes. **Theory of Knowledge.** São Paulo: Ed. Martins Fontes, 2000.

HUMPHREY, C. R., BUTTEL, F.H. Environment, Energy and Society. Belmont: Wadsworth, 1982.

AGRONOMIC INSTITUTE OF CAMPINAS (IAC). Pedological Map of the State of São Paulo: Expanded Legend. Campinas, Embrapa-Solos/IAC, 1999, 64p. and map.

BRAZILIAN INSTITUTE OF GEOGRAPHY AND STATISTICS - IBGE. 2006 Agricultural Census: Brazil, Major Regions and Federative Units. Rio de Janeiro, IBGE, 2009.

INSTITUTO DE PESQUISAS TECNOLÓGICAS DO ESTADO DE SÃO PAULO. Diagnóstico da situação atual dos Recursos Hídricos e estabelecimento de diretrizes técnicas para a elaboração do Plano da Bacia Hidrográfica do Pardo - Relatório Final. São Paulo: IPT/Digeo, 2000a. 255 p. (Technical Report no. 40.670).

JORDÃO, S. **A contribuição da geomorfologia para o conhecimento da fitogeografia nativa do estado de São Paulo e da representatividade das Unidades de Conservação de Proteção Integral. 2011. Thesis (PhD in geography) - University of São Paulo.**

KRONKA, F.J.N. *et. al.* **Inventário florestal da vegetação natural do Estado de São Paulo.** São Paulo: Secretariat for the Environment; Forestry Institute, Official Press, 2005.

LEFBVRE, Henri. **Introduction to Modernity.** Rio de Janeiro: Paz e Terra, 1969.

LEME, S. M. **Compartimentação geomorfológica e organização do espaço em São José do Rio Pardo (SP).** Master's dissertation - University of **São** Paulo, Faculty of Philosophy, Letters and Human Sciences, São Paulo, SP, 1982.

LEOPOLD, L.B.; CLARKE, F.S.; HANSHAW, B. *etal.* **A procedure for evaluating environmental impact** Washington: U. S. Geological Survey, 1971. 13p.

LEOPOLD, Luna. Bergere. **Flood Hydrology and the Floodplain.** In: Coping with the Flood: The Next Phase, White, G.F., and Myers, Mary F., ed. Water Resources Update. Spring issue, 1978.

LIMA, Cleide Rodrigues. **Urbanisation and interventions in the physical environment at the edge of the São Paulo Sedimentary Basin: a geomorphological approach.** Master's dissertation - Faculty of Philosophy, Letters and Human Sciences, University of São Paulo, 1990.

LOMBARDI NETO, Francisco, MOLDENHAUER, W. C. **Rainfall erosivity: its distribution and relationship to soil loss in Campinas, SP.** In: National Research Meeting on Soil Conservation, 3, Recife, 1980.

MANNHEIM, K. **Man and Society in an Age of Reconstruction. Studies in Modem Social Structure (Mensch** und Gesellschaft im Zeiltalter im des Umbaus, Leiden, Holland). Translated by Edward SHILS. New York, Harvest Book, 1940.

MANTOVANI, Juliana da Costa. **Geomorphology Applied to Environmental Impact Studies for road projects: subsidies for Brazilian environmental** bodies sileiros. Master's dissertation. São Paulo: FFLCH/USP, 2015.

MATTOS, D.L. **Lower Mogiana region. Contribution to the study of agrarian geography from the** point of view of land use. **PhD thesis.** FFCEA, USP. São Paulo, 1959.

MEGGERS, Betty J. **Amazonia: man and nature in a counterfeit paradise.** Airlington Heights, Illinois, Harlan Davidson, 1971.

MONTEIRO, C. A. de F. **On an** Index of **Participation of Air Masses and its Possibilities of Application to Climate Classification.** *In:* Revista Geográfica. Rio de Janeiro, v. 33, n. 61, p. 59-69, 1964.

MONTEIRO, C. A. F. **Geosistemas: a história de uma procura.** São Paulo: Contexto, 2000. 127p.

MONTEIRO, C. A. F. **The climate of the southern region.** *In:* CATALDO, D. M. (Org.) Geografia do Brasil, Grande Região Sul. Rio de Janeiro: IBGE, 1963. p. 117-169.

MONTEIRO, C.A.F. **Clima. Greater Southern Region.** Rio de Janeiro: IBGE. v.4, t.1, p-114- 166. 1968.

MORAN, Emílio. **The Human Ecology of Amazonian Populations.** Petrópolis: Vozes, 1990.

MORIN, Edgar. **The well-made head: rethinking reform, rethinking thought.** Rio de Janeiro: Bertrand Brasil: 2000.

MOTA, S.; AQUINO, M. D. **Proposal for an environmental impact assessment matrix.** *In*: VI Italian-Brazilian Symposium on Sanitary and Environmental Engineering. Sanitary and environmental engineering. Vitória-ES. Proceedings... Vitória - ES. 2002.

NAKAMURA, Eduardo Tomio. **Urban interference in surface runoff: aspects of technical criteria in drainage control and urbanisation planning in the Córrego Taboão watershed.** Individual Degree in Geography, Department of Geography, University of São Paulo, 2006.

NIR, D. **Man, a geomorphological agent: an introduction to Anthropic Geomorpholo- gy.** Boston: D. Reidel publishing Co. and Jerusalem: Keter publishing house, 1983. 165 p.

NYE, David. **Electrifying America. Social meanings of a new technology.** Cambria: MIT Press, 1995.

OLIVEIRA, J.B. **Solos do Estado de São Paulo: descrição das classes registradas no mapa pedológico. Campinas, Agronomic Institute, 1999. Scientific Bulletin 45, 112p.**

OLIVEIRA, M.A.F. de. **Geology and Petrography of the São José do Rio Pardo Region, State of São Paulo. PhD Thesis. Institute of Geosciences, USP.** São Paulo, 1972.

OLIVEIRA, R. R. & ENGEMANN, C. **Landscape history and landscapes without history: human presence in the Atlantic Forest of Southeast Brazil.** In: Revista Esboços, Florianópolis, v. 18, n. 25, p. 9-31, Aug. 2011.

PASSOS, Messias Modesto do. **The Dividing Range: geosystem, landscape and eco-history.** Maringá: Eduem, 2006.

PINCHEMEL, P. **La Face de la Terre.** Elements **of Geography.** Paris: Armand Colin, 3ª ed., 1994.

PIRES NETO, A. G. 1992. **The Synthetic-Historical and Analytical-Dynamic Approaches - A Methodological Proposal for Geomorphology. Doctoral Thesis - University of São Paulo. Department of Geography.**

PITTY, A. F. **The Nature of Geomorphology. London, UK: Methuen, 2002.**

QUEIROZ NETO, J. P.; JOURNAUX, A. **Geomorphological Map of São Pedro. Institute of Geography, USP. São Paulo, 1978.**

RODRIGUES, C. **The urbanisation of the metropolis from the perspective of Geomorphology: Tributes to Geographical Readings. In: CARLOS, A. F; OLIVEIRA, A. U. de. (eds.). The Geographies of São Paulo. São Paulo: Contexto, 2004, v. 1, p. 89-114.**

RODRIGUES, C. **Original Morphology and Anthropogenic Morphology in the definition of urban planning spatial units: an example in the São Paulo metropolis.** In: **Revista do Departamento de Geografia, No.** 17, pp. 101-111, 2005.

RODRIGUES, C. **Qualidade Ambiental Urbana: Como Avaliar.** In: Revista do De partamento de **Geografia (USP), São Paulo, v. 1, n. 11, p. 75-85, 1997b.**

ROMARIZ, D. A. **Original Vegetation of the Paraná-Uruguay Basin.** In: **Geographical** Conditions and **Geo-Economic Aspects of the Brazil-Uruguay Basin. Vol. I and II. AGB,** São Paulo, p. 103-110, 1955.

ROSS, J. L. S. **Environmental Geomorphology.** In: **CUNHA, S.B.; GUERRA, A.J.T. (orgs.)**

Geomorfologia do Brasil. 2ª ed., Rio de Janeiro, Bertrand Brasil, 2001, pp. 351- 388.

ROSS, J. L. S.; MOROZ, I. C. **Geomorphological Map of the State of São Paulo. Geomorphology Laboratory. São Paulo: Department of Geography-FFLCH-USP/ Geotechnical Cartography Laboratory - Applied Geology - IPT/FAPESP** (São Paulo State Research **Foundation**), 1997. (Maps and Reports)

RUBIO, Maurício Fava. Geographical Readings of the Process of Appropriation of the Paranapanema River for Electricity Generation. Master's dissertation. São Paulo: FFLCH/USP, 2008.

RUELLAN, F. The Brazilian Shield and Background Folds. University of Brazil. National Faculty of Philosophy, Department of Geography. Rio de Janeiro, 1952.

SANTOS, M. **Espaço e Sociedade**. Petrópolis: Vozes, 1979.

SANTOS, Milton. **The Nature of Space: Technique and Time, Reason and Emotion**. São Paulo: Hucitec, 2ª . Ed.,1996.

SANTOS, Milton. **Thinking the space of man**. 4. ed. São Paulo: Hucitec, 1997.

SAUER, Cari. **The Morphology of Landscape**. *In:* CORRÊA, Roberto Lobato & ROSEN- DAHL, Zeny. Landscape, Time and Culture, 2ª . Ed. Rio de Janeiro: Eduerj, 2004.

SCHEEL-YBERT, R.; DIAS, O.F. Corondó: **Palaeoenvironmental reconstruction and palaeoethnobotanical considerations in a probable locus of early plant culti- vation (south-eastern Brazil)**. *In*: Environmental Archaeology, v.12, p.129-138, 2010.

SCHMITHÚSEN, J. **Die aufgabenkreise der geographischen**. In: Wissenschaft. Ge- ographiche Rundschau, 22(11):431-443, 1970.

SCHNAIBERG, A. **The Environment: From Surplus to Scarcity**. New York: Oxford University Press, 1980.

SELBY, M. **Earth's changing surface: an introduction to geomorphology**. New York: Oxford University Press, 1985.

SETZER, J. **Contribuição para o estudo do clima do Estado de São Paulo**. São Paulo: Salesian

Professional Schools, 1946. 239 p.

SILVEIRA, Maria Laura. **A Geographical Situation: from method to methodology.** In: Território Magazine, No. 6, Jan-Jun, p. 21-28, 1999.

STATE SYSTEM FOR ANALYSING DATA AND STATISTICS (SEADE). **Summary of Data on São Paulo's Municipalities.** Available at www.sead.sp.gov.br. Accessed on 28/01/2014.

SOARES-FILHO, B. S.; CARMO, V.A.; NOGUEIRA, W.J. Methodology for drawing up a chart of the erosion potential of the Rio das Velhas basin (MG). In: Geonomos, v. 6, n.2, p. 45-54, 2010.

TOMMASI, L.R. **Estudo de impacto ambiental.** São Paulo: CETESB: Terragraph Artes e Informática. 1994, 354p.

TOY, T & HADLEY, R. F. **Geomorphology and reclamation of disturbed lands.** London: Academic Press Inc., 1987.

TRICART, J. AND KILIAN, J. **L'éco-géographie.** Librairie François Maspéro, coll. Hé- rodote, 326 p, 1979.

TRICART, J. **Landscape and Ecology.** In: Revista Inter-Facies, No 76, São José do Rio Preto, 1982.

TRICART, J. **Principes et Methodes de la Géomorphologie,** 1976.

Turner II, B.L.; Kasperson, R.E; Meyer, W.B.; Dow, K.M.; Golding, D.; Kaperson, J.X.; Mitchell, R.C.; Ratick, S.J. (1990). **Two types of global environmental change Definitional and spatial-scale issues in their human dimensions.** In: Global Envi- ronmental Change, 1:14-22.

VERSTAPPEN, H.T. **Applied geomorphology: geomorphological surveys for environ- mental development** Amsterdam: Elsevier Scientific Publishing Company, 1983. 473p.

YATSU, E. **Rock Control in Geomorphology.** Sozosha, Tokyo, 1966.

ZUIDAN, R. A. van. **ITC System of geomorphological survey.** Manual ITC Textbook, Enschede, Netherlands, v. 1, chap. 8, 1975.

Printed by Books on Demand GmbH, Norderstedt / Germany